Congenital Anomalies in Iran

Saeed Dastgiri

Congenital Anomalies in Iran

Springer

Saeed Dastgiri
Tabriz Registry of Congenital Anomalies (TRoCA)
School of Medicine, Tabriz University of Medical Sciences
Tabriz, Iran

ISBN 978-981-97-6273-6 ISBN 978-981-97-6274-3 (eBook)
https://doi.org/10.1007/978-981-97-6274-3

This Springer imprint is published by the registered company Springer Nature Singapore Pte Ltd.
The registered company address is: 152 Beach Road, #21-01/04 Gateway East, Singapore 189721, Singapore

This book is dedicated to:
Children and parents in the Tabriz Registry of Congenital Anomalies region for their participation and contribution to this programme over the years.

Acknowledgements

I would like to give my heartfelt thanks and deepest appreciation to my PhD supervisor at the University of Glasgow, Professor David Stone for all his kindness, support, and guidance throughout my study and works in the Glasgow Yorkhill Hospital. What I have learned about the epidemiology and control of congenital anomalies goes originally back to the same place, time, and personnel. Professor Stone often went above and beyond his duty as a supervisor for me throughout the entire process and after.

Setting and growing up the Tabriz Registry of Congenital Anomalies (TRoCA) would not have been possible without the funding and the support of the Iran Ministry of Health, National Public Health Management Centre, and Tabriz University of Medical Sciences.

I extend my sincere gratitude to the children and parents in the region who contributed to, and participated in the TRoCA programme, and to the TRoCA staff for their continued works in the programme for years.

I am grateful to all clients/patients for sharing their experiences with us in the programme, and to health houses, primary care centres, public and private clinics and hospitals, and Iran Welfare and Rehabilitation Organization for participating in our programme.

Introduction

Since 1998, when I was a postgraduate student in the university of Glasgow, United Kingdom, I have closely been in relation and contact with individuals with birth defects, genetic disorders, congenital malformations and disabilities, and families affected by the disruption in the family, enormous psychological impacts and economic catastrophic consequences due to the disorder in one of the family members. That is more than three decades of experiences now.

This book is the story of the works we did in Tabriz Registry of Congenital Anomalies (TRoCA) in Iran, the story of the lessons learnt, and that how we set up a systematic registry using the information and thoroughly documented data provided by the affected individuals and families with genetic and congenital disorders. In the first part, I provide the definition, background information, and an overview of the magnitude, aetiology, prevention, and control approaches of the genetic disorders and congenital anomalies. The second part describes the full story of TRoCA and its goals and managerial structure followed by a discussion about the Epidemiology and Surveillance of Congenital Anomalies in Iran (ESCAI) and Tabriz Foundation for Public Health Genetics (TFPHG). The occurrence pattern and epidemiology of disability and impairments in Iran, and in the TRoCA region are discussed in the third part of the book. In the fourth part, the basic pillars of the community genetics services in Iran, and the health care system of the country in relation to the current prevention services and programmes for the management and community control of congenital anomalies and genetic disorders are reported. I wrote this book for every mother- and father-to-be and for those having an individual with a congenital anomaly in the family, for health care planners, for geneticists and other clinical staff who are involved in the management of genetic and congenital disorders. The fifth part of the book, as the concluding remarks, is for everyone who is interested, or is involved in matters related to the prevention of genetic and congenital disorders.

There is no doubt that all the works at this programme would not have been possible without the support of children and parents in the TRoCA region who participated and contributed to the TRoCA programme for years. I gratefully thank them for the help that they gave me through this long way.

Despite the various rates and occurrence pattern of congenital and genetic anomalies in the globe, the burden is expectedly higher in the low- and middle-income countries due to the fewer resources and services available for affected individuals and families. Therefore, the TRoCA would hopefully be well perceived as a genetic and congenital disorders prevention programme in public health and clinical settings, and by families in the low-recourses areas.

Tabriz Registry of Congenital Anomalies (TRoCA) Saeed Dastgiri
Tabriz, Iran

Contents

Abbreviations[1]

Abnormality or Anomaly	A deviation or departure from what is typical
Amniocentesis	A medical procedure used to remove a small amount of fluid from the sac that surrounds the foetus in the uterus; it is used most often to: (1) diagnose chromosomal or other genetic disorders early in the second trimester of pregnancy; and (2) determine foetal lung maturity before birth
Birth Defects	See Congenital Anomalies
Birth Defects Research and Prevention Society (BDRP)	This organization is a multidisciplinary scientific society founded in 1960 to study the causes and biological processes leading to abnormal development and birth defects at the fundamental and clinical level and appropriate measures for prevention (https://www.birthdefectsresearch.org)
Burden of Disease	A time-based measure that combines years of life lost due to premature mortality and years of life lost due to time lived in states of less than full health
Case	In epidemiological terms, an individual who meets the criteria for inclusion in a surveillance programme. *Note:* Although this term is not commonly used in clinical settings when referring to a patient, it is a term that is widely used in epidemiology
Center for Disease Control and Prevention (CDC)	A leading health protection agency, based in the United States, that collaborates with partners throughout the nation and the world to create expertise, information, and tools that people and communities need to protect

[1] For all the definitions and terminology, the following book was used unless it is indicated in brackets for some terms: World Health Organization. Birth defects surveillance: a manual for programme managers. 2nd ed. Geneva: World Health Organization; 2020.

	their health through health promotion, prevention of disease, injury and disability, and preparedness for new health threats (https://www.cdc.gov)
Chromosomal Abnormality	The excess or absence (whether total or partial) of a chromosome, or structural changes in the chromosome that most commonly produce a set of intellectual and physical problems (congenital anomalies)
Community Genetics Services (CGS)	The art and science of the responsible and realistic application of health and disease-related genetics and genomics knowledge and technologies in human populations and communities to the benefit of individuals therein. Community genetics is multi-, inter-, and transdisciplinary and aims to maximize benefits while minimizing the risk of harm, respecting the autonomy of individuals and ensuring equity (for further details and references, please see section four)
Congenital Anomalies (Birth Defects)	A structural or functional anomaly of organs, systems, or parts of the body that occurs during intrauterine life and is caused by genetic or environmental factors (e.g. exposure to toxic substances, micronutrient deficiencies, or maternal diseases), or both
DALYs	Disability-Adjusted Life Years (https://www.who.int/data/gho/indicator-metadata-registry/imr-details/158)
Disability	A restriction or lack of ability (resulting from an impairment) to perform an activity in the manner or within the range considered normal for a human being
Latin American Collaborative Study of Congenital Malformations (ECLAMC)	This is a clinical and epidemiological investigation programme on the occurrence and causes of congenital anomalies in the Latin American hospital births established in 1967 (http://www.eclamc.org)
Epidemiology	The study of the frequency and distribution of health events and their determinants among human populations, and the application of such research to the prevention and control of health problems
ESCAI	Epidemiology and Surveillance of Congenital Anomalies in Iran
ETOP	Elective Terminations Of Pregnancy
ETOPFA	Elective Termination Of Pregnancy for Foetal Anomaly

European Network of Population-Based Registries for the Epidemiological Surveillance of Congenital Anomalies (EUROCAT)	This is a programme supported by the European Union, represented by the Commission of the European Community for the epidemiological surveillance of congenital anomalies in Europe. The main objectives are to detect and investigate trends in the frequency of congenital anomalies that could be due to the environmental teratogens or mutagens and to evaluate the effectiveness and efficiency of neonatal and perinatal health services. The specific objectives of the EUROCAT registration are: (1) to provide baseline epidemiologic information on congenital anomalies in Europe; (2) to detect and investigate trends in the frequency of congenital anomalies in order to assess the impact of known or suspected risk factors particularly related to exogenic agents, as drugs and the environment, including occupational exposure; (3) to evaluate the effectiveness and efficiency of health services (primary prevention, prenatal diagnosis and termination of pregnancy, treatment); (4) to provide a well-documented data base for aetiologic and clinical research; (5) to act as information centre which could respond to specific needs, such as the assessment of the impact of environmental accidents or change, or the suspicion of teratogenic influences from food, drugs, or other exposures (https://eu-rd-platform.jrc.ec.europa.eu/eurocat)
Foetal Deaths	A foetus that is deceased at delivery; foetal death refers to death prior to the complete expulsion or extraction of a product of conception from its mother, irrespective of the duration of pregnancy; the death is indicated by the fact that, after such separation, the foetus does not breathe or show any other evidence of life, such as beating of the heart, pulsation of the umbilical cord, or definite movement of voluntary muscles
Gestational Age	The time elapsed, measured in weeks, since conception. Because the exact date of conception is not always known, gestational age might also be defined as the time elapsed from the first day of the woman's last normal menstrual cycle. The duration of a normal pregnancy can range from 38 to 42 weeks.

	Gestational age is frequently a source of confusion when calculations are based on menstrual dates; for the purposes of calculation of gestational age from the date of the first day of the last normal menstrual period and the date of delivery, it is borne in mind that the first day is day 0 and not day 1; days 0–6 therefore correspond to "completed week 0"; days 7–13 to "completed week 1"; and the 40th week of actual gestation is synonymous with "completed week 39"; where the date of the last normal menstrual period is not available, gestational age is based on the best clinical estimate; in order to avoid misunderstandings, tabulations are indicated in both weeks and days. (1) Pre-term or premature: less than 37 completed weeks (less than 259 days) of gestation. (2) Term: from 37 completed weeks to less than 42 completed weeks (259 to 293 days) of gestation. (3) Post-term: 42 completed weeks or more (294 days or more) of gestation
Global Burden of Disease (GBD)	GBD study provides a comprehensive picture of mortality and disability across countries, time, age, and sex. It quantifies health loss from hundreds of diseases, injuries, and risk factors, so that health systems can be improved and disparities eliminated (https://www.healthdata.org/research-analysis/gbd)
Incidence	The number of new cases of a disease among a given population and over a given time frame; not used when reporting congenital anomalies (see Prevalence)
International Clearinghouse for Birth Defects Surveillance Research (ICBDSR)	This is a voluntary and non-governmental consortium of 45 population- and hospital-based surveillance programmes from around the world (including Europe, North, Central, and South America, Asia, New Zealand, and Australia) for investigating the occurrence of at least 1 of 39 various congenital anomalies, and collaborating in the prevention programmes, possible causes and adverse outcomes related to birth defects. ICBDSR was established in 1974. The programme is affiliated with the World Health Organization (WHO). The details of the history, size of the population covered, legislation, funding, directors, and case ascertainment can be

	found for each programme on the ICBDSR website (http://www.icbdsr.org)
International Statistical Classification of Diseases and Related Health Problems (ICD)	This is a standard diagnostic classification tool for epidemiology, health management, and clinical purposes in global level (https://icd.who.int/en)
Isolated Anomaly	A single anomaly; most (about 75% in the aggregate) congenital anomalies present as an isolated anomaly. Occasionally, an isolated major anomaly is associated with one or more minor anomalies
Live Birth	The complete expulsion or extraction of a product of conception from a woman's body, irrespective of the duration of the pregnancy, which, after such separation, breathes or shows any other evidence of life such as beating of the heart, pulsation of the umbilical cord, or definite movement of voluntary muscles, whether or not the umbilical cord has been cut or the placenta is attached. Each product of such a birth is considered live born
LMIC	Low- and Middle-Income Countries
Major Congenital Anomaly	A structural change that has significant medical, social, or cosmetic consequences for the affected individual; this type of anomaly typically requires medical intervention
Malformation	A structural defect of an organ, part of an organ, or a larger region of the body that arises during organogenesis (initial formation of a structure). For most organs, organogenesis takes place during the first 8 weeks after fertilization; the resulting structure can be abnormally formed or incompletely formed, or might fail to form altogether. The term is occasionally used, incorrectly, as a synonym for birth defect
MCA	Multiple Congenital Anomalies
Minor Congenital Anomaly	A structural change that poses no significant health problem and tends to have limited social or cosmetic consequences for the affected individual
National Birth Defects Prevention Network (NBDPN)	This is a volunteer-based organization that addresses the issues of birth defects surveillance, research, and prevention under one umbrella by maintaining a national network of state and population-based birth defects programmes (https://www.nbdpn.org)

National Center on Birth Defects and Developmental Disabilities (NCBDDD)	This is an entity located at the CDC in the United States working on the health and well-being of the most vulnerable populations in the country (https://www.cdc.gov/ncbddd)
Neonatal Death	Deaths among liveborn infants during the first 28 completed days of life; neonatal deaths can be subdivided into early neonatal deaths, occurring within the first 7 days of life, and late neonatal deaths, occurring after 7 but before 28 completed days of life; the age at death during the first day of life (day 0) is recorded in units of completed minutes or hours of life; for the second (day 1), third (day 2), and through 27 completed days of life, the age at death is recorded in days
Neonatal Period	The period that commences at birth and ends 28 completed days after birth
Non-Governmental Organization (NGO)	This is a voluntary entity or institution formed independent from the government. Generally, NGOs operating independently with social and moral missions in/for a specified groups in the community
Perinatal Period	The period that commences at 22 completed weeks (154 days) of gestation (the time when birth weight normally is 500 g) and ends 7 completed days after birth
Prenatal Screening	A systematic search for a specific condition among a large, asymptomatic subpopulation of pregnant women selected by personal or family history, or by demographic characteristics such as age and ethnicity; typically, it identifies at-risk groups for further diagnostic testing
Prevalence	A measure of the total number of existing cases of a condition, known as prevalent cases, for a given point in time or period, and among a given population, regardless of whether or not they are new cases; also an indicator of the magnitude of the occurrence of a disease or other health event in the population. In birth defects epidemiology, the terms live birth prevalence and total prevalence are used: (1) Live birth prevalence of congenital anomalies: measures the number of cases with congenital anomalies among live births and is defined as number of cases of live births with any congenital anomaly

	(numerator) among a defined cohort of live births (denominator). For example, the live birth prevalence of congenital anomalies in 2014 is computed as live births born with any congenital anomaly in 2014 divided by all live births born in 2014. (2) Birth prevalence of congenital anomalies: measures the number of cases with congenital anomalies among live births and foetal deaths (stillbirths) and is defined as number of cases of live births and foetal deaths (stillbirths) with any congenital anomaly (numerator) among a defined cohort of live births plus foetal deaths (stillbirths) (denominator). For example, the birth prevalence of congenital anomalies in 2014 is computed as live births plus foetal deaths (stillbirths) with any congenital anomaly in 2014 divided by all live births plus foetal deaths (stillbirths) in 2014. (3) Total prevalence of congenital anomalies: measures the number of cases with congenital anomalies in live births, foetal deaths (stillbirths), plus elective terminations of pregnancy for foetal anomaly, and is defined as number of cases of live births, foetal deaths (stillbirths), elective terminations of pregnancy for foetal anomaly (numerator) among a defined cohort of live births, foetal deaths (stillbirths) and elective terminations (denominator). For example, the total birth prevalence of congenital anomalies in 2014 is computed as live births and foetal deaths (stillbirths) with any congenital anomaly plus elective terminations of pregnancy for foetal anomaly in 2014 (numerator) divided by all live births and foetal deaths (stillbirths) in 2014 plus all elective terminations of pregnancy for foetal anomaly occurring in 2014
Public Health Surveillance	The systematic, continuous, timely, and reliable collection of relevant and necessary data regarding certain health conditions among a population; analysis and interpretation of the data must provide grounds for decision-making and be disseminated
TRoCA	Tabriz Registry of Congenital Anomalies
State Welfare Organization of Iran (SWO)	SWO is a governmental organization established in 1980 in Iran to support homeless

	families, abused and deprived people, orphans, disabled individuals, elderly people, and any other similar vulnerable families and individuals (https://www.behzisti.ir)
Stillbirth	World Health Organization defines stillbirth as the third trimester foetal death (1000 g or more; 28 weeks or more) for international comparison purposes. However, in broader terms, a stillbirth is a foetal death after the gestational age of viability. The definition of viability is based on gestational age and/or weight and is variable among countries
Surveillance Programme	A public health programme that collects, monitors, analyses, interprets, and disseminates data systematically in a timely manner, and that allows for planning, implementation, and evaluation of health strategies
Syndrome	A pattern of multiple anomalies thought to be pathogenetically related and not representing a sequence; it is due to a single cause—genetic or environmental—or to gene–environment interactions
Teratogen	An agent capable of interrupting or altering the normal development of an embryo or foetus, often resulting in a congenital anomaly or embryonic or foetal death
World Health Organization (WHO)	The directing and coordinating authority for health within the United Nations system, responsible for providing leadership on global health matters, shaping the health research agenda, setting norms and standards, articulating evidence-based policy options, providing technical support to countries, and monitoring and assessing health trends
YLD	Years of healthy life lost due to disability (https://www.who.int/data/gho/indicator-metadata-registry/imr-details/160)
YLL	Years of life lost from mortality (https://www.who.int/data/gho/indicator-metadata-registry/imr-details/159)

List of Figures

List of Tables

List of Appendixes

Part I

1 Historical Notes

Contents

1.1 General Background in the World

From the dawn of human history until quite recent decades, in the absence of scientific understanding about the basis of the nature and causes of birth defects, the occurrence of birth defects was believed to result from the action of supernatural forces, hybridization with animals, maternal impressions, celestial bodies, and as the manifestations of evil, and as the sign of displeasure and warning message of admonition from gods. Pregnant women and fathers used to pray to the spiritual world when they were looking forward to the birth of a healthy child [1].

It was generally believed that human embryos were protected from environmental agents such as drugs and viruses by their extraembryonic/foetal membranes and their mothers' abdominal and uterine walls.

The systematic and evidence-based investigations of the occurrence of birth defects began in the mid-twentieth century when Gregg, a researcher from Australia, presented the first well-documented evidence that an environmental agent (rubella virus) could produce severe developmental disruptions such as cataracts if it was present during the critical period of foetus development of the heart, eyes, and ears [2]. The teratogenic effects of exposures to thalidomide was discovered in 1959 when this tranquilizer and antinausea drug caused severe limb and organ abnormalities taken by pregnant women. Later in 1961, Lenz and McBride studied the role of drugs on the occurrence of congenital defects in human [3–5].

The atom bombs dropped on Hiroshima and Nagasaki in the World War showed indubitable evidences indicating that radiation could also be a potent cause of congenital anomalies.

The original and novel works by Johann Gregor Mendel (1822–1884) demonstrated that the heredity characteristics might actually cause some of genetic and congenital disorders. Investigations by Mendel, Morgan, and Avery contributed significantly to understanding the human development. The discovery of the structure of deoxyribonucleic acid (DNA) by Watson and Crick, and the explanation of the genetic codes by Nirenberg and Matthaei, provided the basis for much of the

S. Dastgiri, *Congenital Anomalies in Iran*, https://doi.org/10.1007/978-981-97-6274-3_1

knowledge we have today on the genetic basis of congenital defects in human [1].

In the most recent years, some outbreaks were reported of the disease and complications caused by the Zika virus. This virus is transmitted from the mosquito to humans and usually is common in the tropical regions including parts of the Africa and Southeast Asia. Early this century, the virus did spread out in the areas in Pacific Ocean and America. Some studies have shown that the infection of pregnant women with this virus increased the risk of babies being born with microcephaly. This is a congenital disorder in which the baby's brain is abnormally smaller than the normal size [6].

These discoveries and observations that environmental agents and genetic factors could cause congenital disorders have enormously enhanced our understanding of the scientific basis of the nature and causes of birth defects.

1.2 Hereditary and Genetics in the Iran Medical History

Iranian prophet Zoroaster (~1700 years BC) believed in the equal role of both sexes (men and women) in the formation of a child and its hereditary conditions and characteristics. Similarly, Abul-Qasem Ferdowsi Tusi, one of the Iranian great poets of the tenth century expressed in Shahnameh (one of the world's longest epic poem) the equal contribution of women and men in foetus formation and the transmission of characters to children [7].

Although, it was broadly believed, for centuries, that the congenital anomalies occur by supernatural and superstitious forces, in the beginning of the late first and early second millennium AD, four preeminent physicians and philosophers in Iran argued differently emphasizing the role of inheritance and the effect of forces in the nature on the occurrence of birth defects:

Abu Bakr al-Razi (865–925) was an Iranian physician, philosopher, and chemist. Razi discussed the role of instinct (Seresht) on the development of some diseases in his 23-volume set of medical textbooks called *Al-Kitab al Hawi* [8, 9].

Ali ibn Abbas al-Majusi al-Ahwazi (925–994) was a pioneer Persian Zoroastrian physician mainly worked in the field of psychophysiology and psychosomatic medicine. In the first volume of his *Complete Book of the Medical Art* (*Kitab Kamil a of the Medical Arts*), he says that the disharmony of the body state in patients is caused either by the temperament, shape and natural constitution of the body, or by inheritance due to the background condition of their fathers and ancestors [10, 11].

Abu Ali Abdulloh Ibn-Sina (980–1037), commonly known as Avicenna, was one of the greatest physicians, philosophers, and scientists in Iran history. His most significant work in medicine is a medical encyclopaedia in five volumes called *The Book of Canon in Medicine* (*al-Qānūn fī al-on i*). The Canon used to be a typical common textbook of medicine in many medieval universities for centuries in the world. In the first book of the Canon, Avicenna explained clearly that there are diseases and conditions that are originally hereditary and can be passed from one generation to another. Avicenna wrote that some diseases are specific to a race, an ethnic group, or residents of a region. He classified the hereditary disorders into four major groups in the Canon describing the clinical management of those diseases [12, 13].

The Treasure of Khwarazm Shah (*Zakhireye Khwarazmshahi*) is a medical encyclopaedia in ten volumes compiled by the Iranian physician, Zayn al-Din Sayyed Isma'il ibn Husayn Gorgani, also known as al-Jurjani (1040–1136). He composed a chapter in his book for hereditary diseases that pass from fathers to children, as he argued. He says in the ninth book of his collection that we are excused from suffering from hereditary diseases, but other diseases that occur due to negligence and carelessness are our fault, and we are not therefore excused for this type of disorders [14, 15].

***

In the modern era in Iran, Mohammad Ali Molavi (1921–2016) was a professor of human genetics in Tehran University of Medical Sciences. Dr. Molavi also worked as a member of the World Health Organization expert advisory panel on human genetics. He is known as the pioneer and father of genetics in Iran. Professor Molavi trained many students in the field of genetics and wrote the first textbook of genetics and hereditary disorders in man in three volumes, 1956, in Iran [16].

In 1936, the genetic course was formally included in the curriculum of biology in some of the Iranian universities. Later in 1966, the Iranian Genetics Society was founded.

Professor Dariush Farhud (1938–) founded the first genetic counselling centre in the capital city of Tehran and established the first department of Human Genetics and Anthropology at the University of Tehran in 1972.

Nowadays, there are many professional staff and personnel working in various fields of medical and human genetics in fully equipped centres, departments, and clinics across the country.

***

Until the quite recent years, apart from some occasional publications in the field of some genetic and rare disorders, there was no systematic population data available about the prevalence, aetiology, and surveillance of congenital anomalies in Iran.

Beginning from 2000, a project was carried out on the epidemiology of congenital anomalies aiming to document the epidemiologic features of congenital anomalies in the northwest of Iran and to provide the baseline information to set up the first registry of birth defects in Iran. Accordingly, the Tabriz Registry of Congenital Anomalies (TRoCA) was then established. This is the first programme for the monitoring of congenital defects in the country [17]. The full story of TRoCA in details will be described later in this book.

References

1. Stevenson RE. Causes of human anomalies: an overview and historical perspective. In: Stevenson RE, Hall JG, Goodman HM, editors. Human malformations and related anomalies, vol. 1. New York: Oxford University Press; 1993. p. 3–20.
2. Gregg NM. Congenital cataract following German measles in the mother. Trans Ophtalmol Soc Aust. 1941;3:35.
3. Lenz W. Kindliche missbildungen nach medikament wahrend der graviditat? Dtsch Med Wochenscher. 1961;86:2555.
4. McBride WG. Thalidomide and congenital abnormalities. Lancet. 1961;2:1358.
5. Brent RL, Holmes LB. Clinical and basic science from the thalidomide tragedy; what have we learned about the causes of limb defects? Teratelogy. 1988;38:241.
6. European Centre for Disease Prevention and Control. Rapid risk assessment. Zika virus disease epidemic. Tenth update, 4 April 2017. Stockholm: ECDC; 2017.
7. Behnam B, Zakeri M. Genetics and genomic medicine in Iran. Mol Genet Genomic Med. 2019;7(2):e00606.
8. Adamson P. Abu Bakr al-Razi. The Stanford Encyclopedia of Philosophy (Summer 2021 Edition). Zalta EN, editor. https://plato.stanford.edu/archives/sum2021/entries/abu-bakr-al-razi.
9. Razi M. Al-Kitab al-Hawi. Zaker ME, trans. Tehran: Beheshti University of Medical Sciences; 2015.
10. The NLM History of Medicine Division. [Internet]. Bethesda: National Library of Medicine; 1994 [updated 15 December 2011; cited 29 March 2023]. https://www.nlm.nih.gov/exhibition/islamic_medical/islamic_07.html.
11. Ahvazi A. Kamel al-Sanae al-Tebbie. Moaseseh Ehyae Teb Tabiei: Tehran; 2008.
12. Gutas D. Ibn Sina [Avicenna]. The Stanford Encyclopedia of hilosophy (Fall 2016 Edition). Zalta EN, editor. https://plato.stanford.edu/archives/fall2016/entries/ibn-sina.
13. Avicenna. Al-ghanon fi al-Teb (Canon). Sharfkendi A, trans. Tehran: Sorosh; 2010: Vol. 1.
14. Moattar F, Shams Ardekani MR, Ghannadi A. The life of Jorjani: one of the Persian pioneers of medical encyclopedia compiling. Iran Red Crescent Med J. 2013;15(9):763–6.
15. Jorjani S. Zakhire Kharazmshahi, vol. 2. Tehran: Moaseseh Ehyae Teb Tabiei; 2012.
16. Molavi MA. Genetics and hereditary disorders, vol. 1–3. Tehran: Tehran University Press; 1956.
17. Dastgiri S, Kalankesh LR, Heidarzadeh H, Tajahmad A, Rezaian E. A new registry of congenital anomalies in Iran. J Registry Manag. 2010;37(1):27–9.

Congenital Anomalies at a Glance

2

Contents

This topic discusses the definition, types, and global situation of congenital anomalies in terms of epidemiology and patterns of occurrence. The causes, prevention, and control of birth defects are discussed separately later in this section.

2.1 Definition and Types

The term congenital anomalies is defined as abnormalities in the structure or function of an organ or tissue. Some defects are diagnosed at birth while some others may be detected later at older ages. Anomalies can be isolated or may affect one or more organs at the same time in the body systems. The defects may be categorized as minor or major in terms of clinical importance where some abnormalities have medical, surgical, or cosmetic significance for affected individuals. Minor anomalies are usually seen more frequently than major defects. Congenital anomalies often cause, or are associated with, foetal and infant deaths. They are responsible for a remarkable proportion of foetal deaths including induced abortions, stillbirths, and spontaneous miscarriages. Several thousands of anomalies have already been identified to date. The most common types of defects are:

- Structural malformations (e.g. heart defects, cleft lip, and palate)
- Functional defects (e.g. hearing loss, vision impairments)

S. Dastgiri, *Congenital Anomalies in Iran*, https://doi.org/10.1007/978-981-97-6274-3_2

- Chromosomal anomalies (e.g. Down syndrome)
- Single gene disorders (e.g. cystic fibrosis)
- Multifactorial disorders (e.g. neural tube defects)

For the purposes of the registry and surveillance programmes, the standard coding system of the International Classification of Diseases (ICD) is usually used [1]. The details of the known defects are usually categorized and coded under one of the following main headings:

- nervous system
- eye, ear, face, and neck
- circulatory system
- respiratory system
- cleft lip and cleft palate
- digestive system
- genital organs
- urinary system
- musculoskeletal system
- other congenital malformations
- chromosomal abnormalities

There are also changes in the full list of the anomalies in the new revision of the ICD that are presented in detail in Appendix A.1.

2.2 Prevalence in the Globe

The published reports on the prevalence of congenital anomalies from different countries can be expected to differ due to case ascertainment methods, study definitions of birth defects, data collection, sources of information, type of notification of foetal death, and different time periods over which individuals are ascertained. As these differences can be misleading of comparison, it should therefore be taken into account when international comparisons are made.

The prevalence rates are dependent upon the type of defect groups, and also, the overall occurrence of congenital anomalies varies markedly between different countries and communities ranging approximately from 3% to 7% of births worldwide making them a significant global public health concern. The prevalence and secular trend of groups of congenital anomalies has generally been fluctuating during the recent decades. While the prevalence of neural tube defects, for instance, has been declining in some countries, Down syndrome and anomalies of cleft lip and palate and some other defects were almost stable or had a slight increasing trend over time in some areas. On average, there are approximately 1 in 33 babies born with a birth defect. However, the higher rates usually belong to less developed areas. It is globally estimated that three to eight million children are born with congenital anomalies every year. This accounts for a greater contribution to ill health in childhood and a significant proportion of infant mortality worldwide [2–7].

There have been conflicting reports on the association of the prevalence of birth defects and socioeconomic status and ethnic origins. Total prevalence does not markedly vary among various ethnic and socioeconomic groups [8, 9]. However, the impact of cultural factors, social differences, some dietary deficiencies and hereditary susceptibilities cannot be ignored on the occurrence of different types of birth defects because the social and biological backgrounds may influence the exposures level in different individuals, families, and groups [8–10].

In countries participating in the EUROCAT project of the epidemiological surveillance of congenital anomalies in Europe [11], a total number of 23,893,807 births with 540,556 cases of congenital anomalies were investigated in a period of four decades. The total prevalence rate was 226.2 per 10,000 births ranging from 217 per 10,000 births in 1980 to 253 per 10,000 births in 2020 (Fig. 2.1). The occurrence rate was the highest among infants in the Wales region of the United Kingdom with 368.2 (per 10,000 births) compared to the lowest rate in the Portugal (109.6 per 10,000 births) (Table 2.1) [5].

The most frequent categories of defects were congenital heart diseases (71.4 per 10,000 births), genetic disorders (including skeletal dysplasias, Down syndrome, Patau syndrome, Edwards syndrome, Turner syndrome, triploidy, and polyploidy) with 42.4 (per 10,000 births), and limb

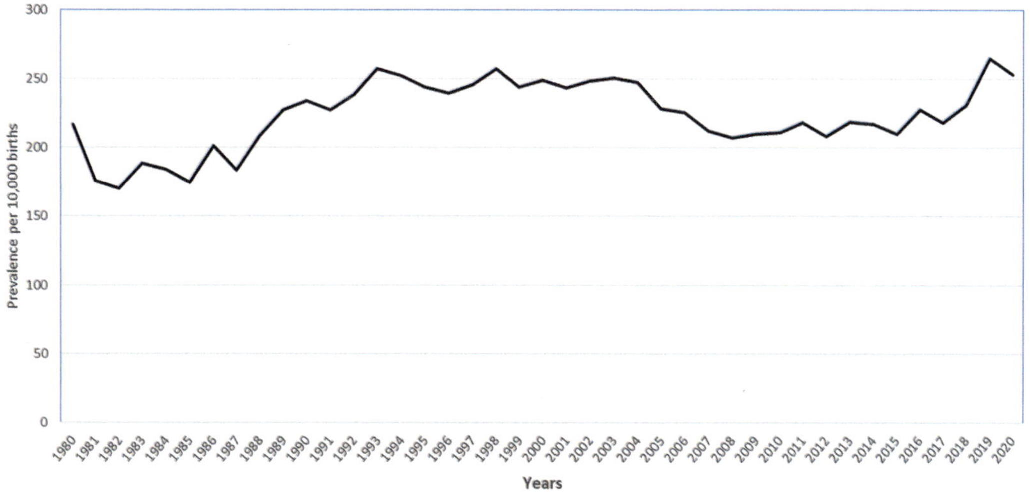

Fig. 2.1 Total prevalence (per 10,000 births) of congenital anomalies in the EUROCAT Network

anomalies (38.3 per 10,000 births). At the same time, the lowest rates belonged to the ear, face, and neck and respiratory anomalies with less than 3 per 10,000 births for each defect. However, prevalence rates do not demonstrate any consistent pattern, and there are remarkable variations between nations over time in the EUROCAT registry regions (Table 2.2) [5].

On the basis of pooled data involving 45 population- and hospital-based surveillance programmes from around the world (including Europe, North, Central and South America, Asia, New Zealand, and Australia), overall rate for 39 selected groups of congenital anomalies were reported by the scientist members of the International Clearinghouse for Birth Defects Surveillance Research (ICBDSR). The centre is an exceptional source to investigate and work on birth defects in the global scale [12]. A summary of the prevalence, trends, mortality, and survival status for selected categories of congenital anomalies is presented in Table 2.3.

The prevalence of anorectal malformation (3.26 per 10,000 total births) decreased over time [13]. Congenital hydrocephalus with total birth prevalence rate of 6.8 per 10,000 births showed an early neonatal case fatality rate of 5.9 per 100 liveborn [14]. Almost all infants with isolated oesophageal atresia survived up to 5 years where the survival improved since 1980, particularly for those with additional anomalies [15]. The prevalence of congenital diaphragmatic hernia slightly increased over time whereas the mortality rates showed a decreasing trend over time [16]. While the overall mortality rate of omphalocoele was 32.1%, one-third of children with this type of defects did not survive early childhood [17]. A high proportion of stillbirth, infant, and child deaths was observed among those with spina bifida. The overall first-week mortality proportion was 6.9% and the majority of the first-week mortality occurred in the first day of life [18]. The live birth prevalence for Trisomy_13 was 0.55, and for Trisomy_18 was 1.07 per 10,000 births. Half of mortality occurred in the first day of life for these defects, and only 10% survived the first year of life after birth [19]. An increasing time trends for the first-, second-, and third-degree hypospadias was observed in a research by ICBDSR [20]. Data from 22 population- and hospital-based registries of birth defects in 18 countries presented the pooled prevalence of 6.4 per 10,000 births for cleft lip with cleft palate. The rate of the terminations of pregnancy for this groups of defects was higher in the most European countries compared to other continents. An overall survival rate of 91% was reported for total cleft lip with cleft palate whereas the similar rate was 77.1% for the cases with multiple congenital anomalies [21].

Table 2.1 Prevalence and termination of pregnancy for birth defects (per 10,000 births) in the EUROCAT Network between 1980 and 2020 by Country

Registry	Total prevalence	Termination of pregnancy for fetal anomaly
Portugal—S Portugal	109.63	16.52
Poland	141.7	0.02
Croatia—Zagreb	181.81	10.33
Ireland—SE Ireland	188.98	1.89
United Kingdom—East Midlands & South Yorkshire	211.39	44.11
United Kingdom—Wessex	211.66	58.92
Italy—Emilia Romagna	212.47	31.52
Italy—Tuscany	213.79	38.73
United Kingdom—Thames Valley	219.55	59.34
Spain—Basque Country	224.86	65.03
Bulgaria—Pleven	229.64	46.54
United Kingdom—West Midlands	235.87	51.88
Belgium—Hainaut	239.66	36.02
Spain—Valencian Region	242.89	56.8
Ireland—Cork and Kerry	243.65	8.99
France—French West Indies	244.19	72.82
Ukraine	246.52	35.01
United Kingdom—Yorkshire and Humber	247.68	50.06
United Kingdom—Northern England	248.44	53.24
Poland—Wielkopolska	253.74	0.08
Denmark—Funen	259.92	37.94
NL—N Netherlands	263.12	25.21
Belgium—Antwerp	264.51	29.59
Austria—Styria	269.41	30.1
Malta	282.17	0
Italy—Milan Area	283.17	54.09
United Kingdom—South West England	285.15	59.51
Germany—Saxony-Anhalt	289.78	33.23
Norway	298.46	36.07
France—Paris	298.59	80.25
France—Isle de la Reunion	300.87	59.95
Spain—Navarre	304.86	84.89
Italy—Trento	307.54	82.15
France—Auvergne	330.98	68.19
Switzerland—Vaud	364.96	68.16
France—Brittany	366.39	79.88
United Kingdom—Wales	368.23	52.89

The most recent research figures indicate that although the global health has improved over the past few decades as measured by age-standardized DALY rates in many countries and territories, congenital birth defects were similarly ranked fourth with 2% increase from 1990 to 2019 among the causes of DALYs in children younger than 10 years between two period of time: 6.6% of DALYs and 8.6% of DALYs for 1990 and 2019, respectively [22].

It is generally supposed that the occurrence of birth defects might be higher in less developed countries compared to the developing areas because there are usually lower access to the expensive medical procedures and genetics technology for screening, diagnosis, and treatment of

Table 2.2 Prevalence and termination of pregnancy for birth defects (per 10,000 births) in the EUROCAT Network between 1980 and 2020 by Selected Groups of Anomalies

Anomaly group	Total prevalence	Termination of pregnancy for fetal anomaly
All anomalies	226.23	32.89
Nervous system anomalies	21.96	9.23
– Neural tube defects	8.92	5.42
– Anencephaly and similar	3.17	2.59
– Encephalocele and meningocele	1.05	0.59
– Spina Bifida	4.7	2.24
– Hydrocephaly	4.78	1.74
– Severe microcephaly	2.24	0.21
– Arhinencephaly / holoprosencephaly	1.15	0.79
– Agenesis of corpus callosum	1.32	0.46
Eye anomalies	3.86	0.26
– Anophthalmos / microphthalmos	0.94	0.16
– Anophthalmos	0.21	0.05
– Congenital cataract	1.02	0.01
– Congenital glaucoma	0.23	0
Ear, face and neck anomalies	2.87	0.29
– Anotia and atresia / stenosis / stricture of external auditory canal	0.97	0.04
Congenital heart defects	71.36	5.51
– Severe congenital heart defects	18.32	3.07
– Common arterial truncus	0.69	0.18
– Double outlet right ventricle	1.07	0.24
– Double outlet left ventricle	0.05	0.02
– Complete transposition of great arteries (D-TGA)	3.02	0.28
– Single ventricle	0.67	0.21
– Corrected transposition of great arteries (L-TGA)	0.24	0.03
– Ventricular septal defect	33.12	1.53
– Atrial septal defect	17.29	0.36
– Atrioventricular septal defect	3.97	0.86
– Tetralogy and pentatology of Fallot	3.06	0.34
– Triscuspid atresia and stenosis	0.66	0.15
– Ebstein's anomaly	0.41	0.05
– Pulmonary valve stenosis	3.64	0.09
– Pulmonary valve atresia	0.94	0.18
– Aortic valve atresia/stenosis	1.22	0.09
– Mitral valve atresia/stenosis	0.45	0.1
– Hypoplastic left hear (HLH/HLHS)	2.38	0.88
– Hypoplastic right heart (HRH/HRHS)	0.39	0.14
– Coarctation of aorta	3.29	0.18
– Aortic atresia / interrupted aortic arch	0.36	0.07
– Total anomalous pulmonary venous return	0.5	0.03
– PDA as only CHD in term infants	2.58	0
Respiratory anomalies	2.89	0.47
– Choanal stenosis or atresia	0.73	0.03
– Congenital pulmonary airway malformations (CPAM)	0.68	0.06
Oro-facial clefts	14.46	1.12
– Cleft lip with or without cleft palate	8.65	0.76
– Cleft palate	5.81	0.36

(continued)

Table 2.2 (continued)

Anomaly group	Total prevalence	Termination of pregnancy for fetal anomaly
Gastro-intestinal anomalies	16.47	1.82
– Oesophageal atresia with or without trachea-oesophageal fistula	2.31	0.15
– Duodenal atresia or stenosis	1.2	0.08
– Atresia or stenosis of other parts of small intestine	0.65	0.02
– Ano-rectal atresia orand stenosis	2.91	0.47
– Hirschsprung's disease	1.02	0
– Atresia of bile ducts	0.26	0.01
– Annular pancreas	0.16	0.02
– Anomalies of intestinal fixation	1.17	0.31
– Diaphragmatic hernia	2.5	0.49
Abdominal wall defects	5.34	1.91
– Gastroschisis	2.2	0.28
– Omphalocele	2.82	1.39
Congenital anomalies of kidney and urinary tract (CAKUT)	29.86	3.53
– Unilateral renal agenesis	3.09	0.38
– Bilateral renal agenesis including Potter sequence	1.17	0.69
– Multicystic renal dysplasia	3.04	0.55
– Congenital hydronephrosis including ureter obstruction	12.43	0.45
– Lobulated, fused and horseshoe kidney and ectopic kidney	2.14	0.38
– Bladder exstrophy and / or epispadia	0.57	0.11
– Posterior urethral valve	0.86	0.12
– Prune belly syndrome	0.15	0.08
Genital anomalies	18.5	0.57
– Hypospadias	15.13	0.08
– Indeterminate sex	0.58	0.1
Limb anomalies	38.28	3.29
– Limb reduction defects (LRD)	5.24	1.12
– Transverse LRD	0.66	0.18
– Longitudinal preaxial LRD	0.91	0.35
– Longitudinal postaxial LRD	0.28	0.1
– Longitudinal central LRD	0.29	0.06
– Intercalary LRD	0.13	0.06
– Club foot—talipes equinovarus	9.26	0.72
– Hip dislocation	5.07	0.01
– Polydactyly	9.48	0.49
– Syndactyly	4.87	0.3
Other anomalies/syndromes		
– Craniosynostosis	1.93	0.09
– Congenital constriction bands / amniotic band sequence resulting in major malformations	0.38	0.15
– Situs inversus	0.62	0.12
– Conjoined twins	0.14	0.1
– VATER / VACTERL association	0.34	0.07
– Pierre Robin sequence	0.95	0.04
– Caudal regression sequence	0.08	0.04
– Sirenomelia	0.07	0.05
– Septo-optic dysplasia	0.1	0.01
– Vascular disruption anomalies	4.43	0.69
– Laterality anomalies	1.65	0.39

(continued)

Table 2.2 (continued)

Anomaly group	Total prevalence	Termination of pregnancy for fetal anomaly
– Teratogenic syndromes resulting in major malformations	0.6	0.11
– Valproate syndrome	0.03	0.01
– Maternal infections resulting in major malformations	0.37	0.11
Genetic disorders	42.43	18.04
– Skeletal dysplasias	1.52	0.63
– Down syndrome / trisomy 21	18.88	8.72
– Patau syndrome / trisomy 13	1.57	1.13
– Edwards syndrome / trisomy 18	4.07	2.98
– Turner syndrome	1.98	1.22
– Triploidy and polyploidy	0.67	0.57

Table 2.3 Prevalence, mortality and survival for selected groups of congenital anomalies in the ICBDSR Network of Global Registries

Anomaly type	Main findings
Anorectal malformation [13]	Prevalence was 3.26 (per 10,000 births)
	60% of cases were multiple or syndromic
	The prevalence rates decreased over time
	First week mortality proportion was 12.5%
Congenital hydrocephalus [14]	Prevalence rate was 6.8 (per 10,000 births)
	Early neonatal case fatality rate was 5.9 (per 100 liveborn cases)
	Case fatality rate was 2.7 times higher in syndromic cases compared to non-syndromic cases
Esophageal atresia [15]	Survival rate was 89.4% at 1-month, 84.5% at 1-year and 82.7% at 5-years
	Almost all infants with isolated esophageal atresia survived to 5 years
	Survival status improved markedly from the 1980s
Congenital diaphragmatic hernia [16]	Prevalence was 2.6 (per 10,000 births)
	The prevalence rates slightly increased over time
	Total mortality rate was 37.7%
	The mortality rates decreased over time
	The most deaths occurred in 2–6 days old infants
Omphalocele [17]	Prevalence was 2.6 (per 10,000 births)
	Mortality rate was 32.1%
	One-third of children did not survive early childhood
Spina bifida [18]	First-week mortality was 6.9%
	Majority of first-week mortality occurred on the first day of life
	The rates of stillbirth, infant mortality and child deaths were high
Trisomy 13 and trisomy 18 [19]	Prevalence for T13 was 0.55 (per 10,000 births)
	Prevalence for T18 was 1.07 (per 10,000 births)
	Median mortality in the first week of life was 48% for T13 and 42% for T18
	Mortality in the first year of life was 87% for T13 and 88% for T18
	Half of mortality occurred on the first day of life
	10% survived to the first year of life
Hypospadias [20]	Prevalence was 20.9 (per 10,000 births)
	The first, second and third degree anomalies increased over time
Cleft lip with cleft palate [21]	Prevalence was 6.4 (per 10,000 births)
	Terminations of pregnancy for this anomaly was higher in the most European countries compared to other continents
	The total survival rate was 91%
	The survival rate was 77.1% for the cases with multiple defects

congenital anomalies. Therefore, there is expectedly a paucity of evidences on the true figure of congenital anomalies in the population level in resource-limited settings.

2.3 Prevalence in the Developing Countries

Research from southwestern Ethiopia reported a prevalence rate of 55 per 10,000 births mainly including anencephaly (25.0%), hydrocephalus (24.6%), spina bifida (13.1%), meningomyelocele (7.1%), and umbilical hernia (4.8%) accounting for about 75% of all recorded birth defects in the region [23].

Another study from the same country reported a prevalence of 62 per 10,000 births [24]. A various prevalence of 190 per 10,000 births was also reported from central and northwest of Ethiopia in a period of 2010–2014 [25]. Another data obtained from newborns admitted in a tertiary hospital in Enugu, Nigeria, presented a prevalence of 280 per 10,000 births mainly surgical defects including cleft lip/cleft palate, neural tube defects, limb abnormalities, omphalocoele, umbilical herniae, anorectal malformations and dysmorphism associated with multiple congenital abnormalities [26]. Similar investigation from Nigeria reported a prevalence rate of 630 per 10,000 births in which the cardiovascular and digestive system anomalies were the most common ones [27]. The data collection in all those studies were hospital based.

There is also an ongoing project entitled "Sub-Saharan African Network for Congenital Anomalies: Surveillance, Prevention and Care" supported and leading by the MU-JHU Care Ltd. [28]. This programme has estimated that 10% of child deaths in the under 5 years of age group are due to the congenital anomalies in the region [29].

2.4 Prevalence in Iran

There is no long-term investigation on the epidemiology and secular trend of congenital anomalies in Iran. Various prevalence rates of congenital anomalies have been reported from different parts of the country.

On the basis of pooled data involving 297,734 births registered by the Tabriz Registry of Congenital Anomalies (TRoCA) in the northwest region of Iran between 2000 and 2020, an overall prevalence rate of congenital anomalies was estimated as 262.9 per 10,000 births. Anomalies of genitourinary tract (59.2 per 10,000 births), limb defects (46.3 per 10,000 births), nervous system anomalies (39.4 per 10,000 births), gastrointestinal anomalies (27.8 per 10,000 births), and congenital heart diseases (37.1 per 10,000 births) accounted for 80% of birth defects. The lowest prevalence rates were reported for musculoskeletal and connective tissue, cleft lip with/without palate, chromosomal, eye and ear anomalies all of which collectively accounted for less than 20 per 10,000 births. Over the past two decades, the region has experienced a highly increasing trend in the prevalence of birth defects. The total prevalence of congenital anomalies was more than tripled in the region over the past two decades ranging from 104.6 (per 10,000 births) in 2000 to 744 (per 10,000 births) in 2020 [30].

A review article based on the 27 studies (1992–2014) was conducted to estimate the occurrence of congenital anomalies among infants in Iran giving a prevalence rate of 230 per 10,000 births. According to this systematic review, the highest prevalence rates belonged to musculoskeletal anomalies, skin anomalies, and genitourinary system anomalies. Prevalence for respiratory system defects was the lowest rate compared to other main groups of anomalies [31].

2.5 Survival and Prognosis

In contrast to the wide range of published research reports on the prevalence of congenital anomalies, a very few published data are available about the survival of birth defects.

A follow-up study in the United Kingdom showed that nearly 90% of infants born with a defect survive up to 5 years. There are, however, variations in the survival rates among different types of birth defects. The prognosis was poorer for children with chromosomal anomalies (excluding Down syndrome) compared to neural tube defects, respiratory system anomalies, congenital heart disease, nervous system anomalies, and Down syndrome [32].

A retrospective study of selected types of congenital anomalies in the Czech Republic showed that the survival in the first year of life was 63% for congenital hydrocephalus, 75–84% in neural tube defects, and 82% to 91% for the defects of abdominal wall, diaphragmatic hernia, congenital defects of oesophagus, and anorectal malformations [33].

The survival rates of up to 20 years of age for some selected groups of congenital anomaly were estimated by the scientists in the Newcastle University of the United Kingdom. The overall 20-year survival rate for all birth defects was 85.5%, 89.5% for cardiovascular system anomalies, 79.1% for chromosomal anomalies, 93.2% for urinary system anomalies, 83.2% for digestive system anomalies, 97.6% for orofacial clefts, and 66.2% for nervous system anomalies [34].

Another study in the New South Wales region of the Australia indicated that half of deaths occurred in the first week of life. The 5-year survival rate was 92.8%. The similar rate was 83.2% for syndromes, 83.4% for multiple anomalies, 85.1% for chromosomal, and 95.3% for isolated defects. Multiple anomalies had poorer prognosis compared to the chromosomal and syndromic disorders [35].

Norwegian cohort analysis in a long-term study of survival of birth defects among female subjects at the Medical Birth Registry of Norway showed that 80% survived up to 15 years of age [36].

A recent research by the ICBDSR affiliated registry members showed a total survival rate of 91% for cleft lip with cleft palate. The survival for isolated cases was 97.7%. Cases of cleft lip with cleft palate associated with multiple congenital anomalies had an overall survival of 77.1% while cases associated with genetic or chromosomal syndromes had a survival of 40.9% [21].

Long-term survival of children born with congenital anomalies was assessed in a recent systematic review and meta-analysis of population-based studies [37]. Accordingly, the 5-year survival for all types of congenital anomalies varied between 85% and 95% in different studies across the counties and ethnic groups in the world. Survival was generally lower for neural tube defects where the survival of congenital heart disease was relatively high compared to other lethal anomalies. Table 2.4 was abstracted from this systematic review to show the overall survival estimates by groups of congenital anomaly type at age 1–25 years. The overall pattern of survival status has considerably improved over the two decades, namely for children with spina bifida, oesophageal atresia, biliary atresia, diaphragmatic hernia, gastroschisis, and Down syndrome. Table 2.5 shows the survival rates for selected groups of congenital anomalies as compared between those born in 2000 and 2020.

Table 2.4 Overall survival rates of congenital anomalies [36]

All congenital anomalies	Deaths/live births	1 year	5 years	10 years	15 years	20 years	25 years
International Classification of Diseases Revision 9 (ICD-9) codes 740.0–759.9	3620/45,200	93.4	92.5	92.3	–	–	–
ICD-9 codes 740–759	2182/43,708	95.7	95.0	–	–	–	–
British Paediatric Association (BPA)-ICD-9 codes 740–759	740/6153	89.1	87.9	–	–	–	–
ICD-8 codes (740–759)	1169/9186	–	–	–	–	87.4	–
ICD-9 740.00–758.090	3518/48,391	93.7	92.7	–	–	–	–
ICD-10 (Q00-Q99)	1465/10,850	–	–	–	–	85.5	–
ICD-9 codes 740–759	9112/57,002	87.1	85.2	–	83.9	–	82.7

Table 2.5 Survival rates for selected groups of congenital anomalies born in 2000 and 2020 [36]

All congenital anomalies	Birth year	1 year	5 years	10 years	20 years	25 years
Spina bifida	Born in 2000	88	87	86	82	81
	Born in 2020	93	92	91	89	88
Encephalocele	Born in 2000	73	73	73	72	72
	Born in 2020	73	72	72	71	71
Oesophageal atresia	Born in 2000	86	86	85	85	84
	Born in 2020	93	93	93	92	92
Biliary atresia	Born in 2000	87	85	82	73	–
	Born in 2020	95	94	92	88	–
Diaphragmatic hernia	Born in 2000	67	67	67	66	66
	Born in 2020	84	83	83	83	83
Gastroschisis	Born in 2000	90	90	89	88	–
	Born in 2020	94	93	93	92	–
Down syndrome	Born in 2000	97	96	96	95	–
	Born in 2020	98	97	97	96	–

2.6 Predisposing Factors Associated with Survival and Prognosis

The literature on the congenital anomalies indicates that the prognosis and survival pattern of congenital anomalies varies by some factors and different communities. However, the overall pattern of survival in children with defects is usually poorer when compared to the general populations [32, 38–40]. Most studies indicate that the presence of extra defects is an important predisposing factor to reduce the survival rate in children with congenital anomalies [41–46].

Ethnicity, gender, socioeconomic status, maternal age, preterm birth, birth weight and healthcare programmes have been investigated in relation to the survival pattern of children born with birth defects [32, 37, 42–44, 47–51]. The survival rates of congenital anomalies are generally lower among males than females, and in subjects with birth weight less than 2500 g and preterm births. There have, however, been inconsistent reports about the impact of ethnicity and socioeconomic status on the prognosis and survival pattern of the children born with congenital anomalies and the subgroups.

2.7 Detection and Termination of Pregnancy

Figure 2.2 presents the prevalence (per 10,000 births) of foetal deaths/still births (from 20 weeks gestation) and termination of pregnancy for foetal anomaly following prenatal diagnosis in the EUROCAT Network. More than 46% of birth defects are prenatally diagnosed (i.e. ultrasound and various screening methods) in Europe ranging from 27% in Austria (Styria) to 69% in the French West Indies [52]. And 44.3 of pregnancies (per 10,000 births) are terminated following prenatal diagnosis for congenital anomaly ranging from 2.1 (per 10,000 births) in 1980 to 58.2 (per 10,000 births) in 2020 among the countries affiliated to the EUROCAT network of birth defects while, at the same period of time and place, the foetal deaths/still births (from 20 weeks gestation) decreased from 7.3 (per 10,000 births) to 5.1 (per 10,000 births). Malta and Poland experienced the lowest rate of the termination of pregnancy for congenital anomalies whereas Spain had the highest rate in the EUROCAT at the same time period. Similarly, Poland and Bulgaria had the lowest and highest still births rate (from 20 weeks gestation), respectively. Studies on spe-

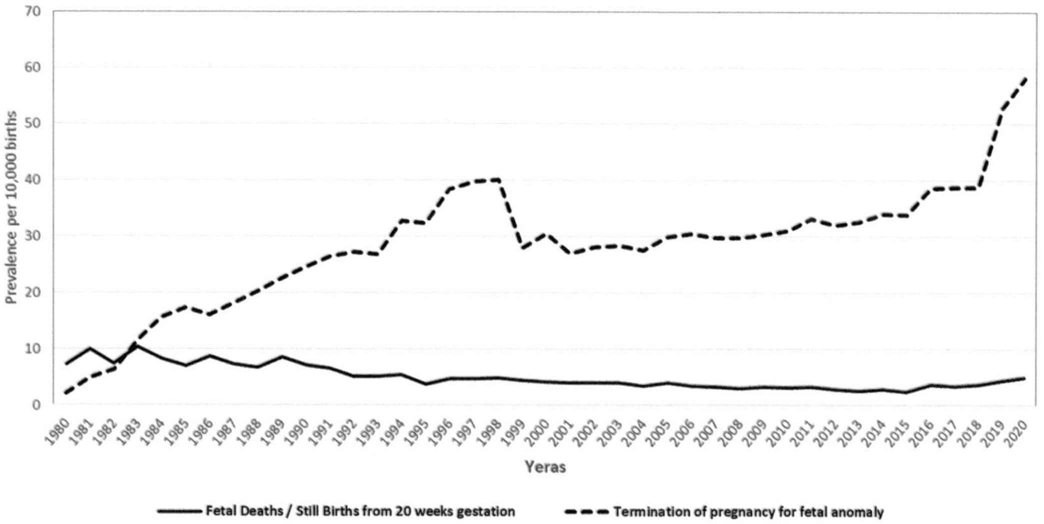

Fig. 2.2 Prevalence (per 10,000 births) of foetal deaths/still births and termination of pregnancy for foetal anomaly in the EUROCAT Network

cific groups of congenital anomalies indicated that 18.1 (per 10,000 births) of genetic disorders (including skeletal dysplasias, Down syndrome, Patau syndrome, Edwards syndrome, Turner syndrome, triploidy, and polyploidy) are terminated after prenatal diagnosis followed by the nervous system anomalies (9.2 per 10,000 births). Termination of pregnancy was also decided in 8.4 (per 10,000 births) of cases with congenital heart defects [5]. A research investigation in Iran reported that the neural tube defects, limb deformation, hydrops fetalis, hydrocephaly, and chromosomal anomalies (including Down syndrome) accounted proportionally for about 65% of defects eligible for abortion [53].

The total rate of foetal death, early neonatal death and perinatal mortality associated with birth defects have been reported 4.5 (per 10,000 births), 4.7 (per 10,000 births) and 9.2 (per 10,000 births), respectively, in European countries. Nervous system defects and chromosomal anomalies showed the highest rates of foetal death due to the congenital anomalies. The similar figures for the early neonatal death belonged to the congenital heart defects and urinary tract anomalies [54].

References

1. World Health Organization. International classification of diseases (ICD). 2023. https://icd.who.int/en.
2. Christianson A, Howson CP, Modell B. March of Dimes global report on birth defects: the hidden toll of dying and disabled children. March of Dimes birth defects foundation; 2006. https://dev.marchofdimes.org/materials/global-report-on-birth-defects-the-hidden-toll-of%2D%2Dd2unzZI5_VWOaLZnw6iHcx7hbpMWtWzTuIOU3DabcVY.pdf.
3. National Birth Defects Prevention Network. Major birth defects data from population-based birth defects surveillance programs in the United States, 2006–2016. Birth Defects Res. 2020;112(18):1424–36.
4. International Clearinghouse for Birth Defects Surveillance Research (ICBDSR). Birth Defects Prevalence. 2023. http://www.icbdsr.org/resources/annual-report.
5. European network of registries for the epidemiologic surveillance of congenital anomalies (EUROCAT). 2023. https://eu-rd-platform.jrc.ec.europa.eu/eurocat/eurocat-data/prevalence_en.
6. UpToDate. Birth defects: epidemiology, types, and patterns. 2023. https://www.uptodate.com/contents/birth-defects-epidemiologytypes-and-patterns.
7. Kouame BD, N'guetta-Brou IA, Kouame GS, Sounkere M, Koffi M, Yaokreh JB, Odehouri-Koudou T, Tembely S, Dieth GA, Ouattara O, Dick R. Epidemiology of congenital abnormalities in West Africa: results of a descriptive study in teaching hospitals in Abidjan: Cote d'Ivoire. Afr J Paediatr Surg. 2015;12(1):51–5.

8. Egbe A, Lee S, Ho D, Uppu S. Effect of race on the prevalence of congenital malformations among newborns in the United States. Ethn Dis. 2015;25:226.
9. Aggarwal D, Warmerdam B, Wyatt K, et al. Prevalence of birth defects among American-Indian births in California, 1983–2010. Birth Defects Res A Clin Mol Teratol. 2015;103:105.
10. Egbe AC. Birth defects in the newborn population: race and ethnicity. Pediatr Neonatol. 2015;56:183.
11. European network of registries for the epidemiologic surveillance of congenital anomalies (EUROCAT). 2023. https://eu-rd-platform.jrc.ec.europa.eu/eurocat.
12. International Clearinghouse for Birth Defects Surveilance Reseach (ICBDSR). 2023. http://www.icbdsr.org.
13. Kancherla V, Sundar M, Tandaki L, Lux A, Bakker MK, Bergman JE, et al. Prevalence and mortality among children with anorectal malformation: a multi-country analysis. Birth Defects Res. 2023;115(3):390–404.
14. Gili JA, López-Camelo JS, Nembhard WN, Bakker M, de Walle HEK, Stallings EB, et al. Analysis of early neonatal case fatality rate among newborns with congenital hydrocephalus, a 2000–2014 multi-country registry-based study. Birth Defects Res. 2022;114(12):631–44.
15. Bell JC, Baynam G, Bergman JEH, Bermejo-Sánchez E, Botto LD, Canfield MA, et al. Survival of infants born with esophageal atresia among 24 international birth defects surveillance programs. Birth Defects Res. 2021;113(12):945–57.
16. Politis MD, Bermejo-Sánchez E, Canfield MA, Contiero P, Cragan JD, Dastgiri S, et al. International Clearinghouse for Birth Defects Surveillance and Research. Prevalence and mortality in children with congenital diaphragmatic hernia: a multicountry study. Ann Epidemiol. 2021;56:61–69.e3.
17. Nembhard WN, Bergman JEH, Politis MD, Arteaga-Vázquez J, Bermejo-Sánchez E, Canfield MA, et al. A multi-country study of prevalence and early childhood mortality among children with omphalocele. Birth Defects Res. 2020;112(20):1787–801.
18. Bakker MK, Kancherla V, Canfield MA, Bermejo-Sanchez E, Cragan JD, Dastgiri S, et al. Analysis of mortality among neonates and children with Spina bifida: an international registry-based study, 2001–2012. Paediatr Perinat Epidemiol. 2019;33(6):436–48.
19. Goel N, Morris JK, Tucker D, de Walle HEK, Bakker MK, Kancherla V, et al. Trisomy 13 and 18-prevalence and mortality-a multi-registry population based analysis. Am J Med Genet A. 2019;179(12):2382–92.
20. Yu X, Nassar N, Mastroiacovo P, Canfield M, Groisman B, Bermejo-Sánchez E, Ritvanen A, et al. Hypospadias prevalence and trends in international birth defect surveillance systems, 1980–2010. Eur Urol. 2019;76(4):482–90.
21. Mc Goldrick N, Revie G, Groisman B, Hurtado-Villa P, Sipek A, Khoshnood B, et al. A multi-program analysis of cleft lip with cleft palate prevalence and mortality using data from 22 international clearinghouse for birth defects surveillance and Research programs, 1974–2014. Birth Defects Res. 2023;115:980–97.
22. GBD 2019 Diseases and Injuries Collaborators. Global burden of 369 diseases and injuries in 204 countries and territories, 1990-2019: a systematic analysis for the Global Burden of Disease Study 2019. Lancet. 2020;396(10258):1204–22.
23. Geneti SA, Dimsu GG, Sori DA, Amente LD, Kurmane ZM. Prevalence and patterns of birth defects among newborns in southwestern Ethiopia: a retrospective study. Pan Afr Med J. 2021;40:248.
24. Mekonnen D, MollaTaye, Worku W. Congenital anomalies among newborn babies in Felege-Hiwot Comprehensive Specialized Referral Hospital, Bahir Dar, Ethiopia. Sci Rep. 2021;11(1):11027.
25. Taye M, Afework M, Fantaye W, Diro E, Worku A. Magnitude of birth defects in central and Northwest Ethiopia from 2010-2014: a descriptive retrospective study. PLoS One. 2016;11(10):e0161998.
26. Obu HA, Chinawa JM, Uleanya ND, Adimora GN, Obi IE. Congenital malformations among newborns admitted in the neonatal unit of a tertiary hospital in Enugu, South-East Nigeria—a retrospective study. BMC Res Notes. 2012;5:177.
27. Ajao AE, Adeoye IA. Prevalence, risk factors and outcome of congenital anomalies among neonatal admissions in OGBOMOSO. Nigeria BMC Pediatr. 2019;19(1):88.
28. UK Research and Innovation (UKRI). 2023. https://gtr.ukri.org.
29. Sub-Saharan African Network for Congenital Anomalies: Surveillance, Prevention and Care. 2023. https://gtr.ukri.org/projects?ref=MRpercent2FT039132percent2F1#/tabOverview.
30. Tarighat F, Golshan E, Dastgiri S. Prevalence of congenital anomalies in the Northwest of Iran. Depiction Health. 2021;12(4):417–25. (Persian).
31. Vatankhah S, Jalilvand M, Sarkhosh S, Azarmi M, Mohseni M. Prevalence of congenital anomalies in Iran: a review article. Iran J Public Health. 2017;46(6):733–43.
32. Dastgiri S, Gilmour WH, Stone DH. Survival of children born with congenital anomalies. Arch Dis Child. 2003;88(5):391–4.
33. Sípek A, Gregor V, Horácek J, Masátová D. Prezívání detí narozených s vybranými typy vrozených vad v Ceské republice [Survival of children born with selected types of birth defects in Czech Republic]. Ceska Gynekol. 2004;69 Suppl 1:47–52. (Czech).
34. Tennant PW, Pearce MS, Bythell M, Rankin J. 20-year survival of children born with congenital anomalies: a population-based study. Lancet. 2010;375(9715):649–56.
35. Schneuer FJ, Bell JC, Shand AW, Walker K, Badawi N, Nassar N. Five-year survival of infants with major congenital anomalies: a registry based study. Acta Paediatr. 2019;108(11):2008–18.
36. Skjaerven R, Wilcox AJ, Lie RT. A population-based study of survival and childbearing among female sub-

jects with birth defects and the risk of recurrence in their children. N Engl J Med. 1999;340(14):1057–62.
37. Glinianaia SV, Morris JK, Best KE, Santoro M, Coi A, Armaroli A, Rankin J. Long-term survival of children born with congenital anomalies: a systematic review and meta-analysis of population-based studies. PLoS Med. 2020;17(9):e1003356.
38. Agha MM, Williams JI, Marrett L, To T, Dodds L. Determinants of survival in children with congenital abnormalities: a long-term population-based cohort study. Birth Defects Res A Clin Mol Teratol. 2006;76(1):46–54.
39. Eide MG, Skjaerven R, Irgens LM, Bjerkedal T, Oyen N. Associations of birth defects with adult intellectual performance, disability and mortality: population-based cohort study. Pediatr Res. 2006;59(6):848–53.
40. Berger KH, Zhu BP, Copeland G. Mortality throughout early childhood for Michigan children born with congenital anomalies, 1992–1998. Birth Defects Res A Clin Mol Teratol. 2003;67(9):656–61.
41. Frid C, Drott P, Lundell B, Rasmussen F, Anneren G. Mortality in down syndrome in relation to congenital malformations. J Intellect Disabil Res. 1999;43(Pt 3):234–41.
42. Rankin J, Tennant PW, Bythell M, Pearce MS. Predictors of survival in children born with down syndrome: a registry-based study. Pediatrics. 2012;129(6):e1373–81.
43. Glinianaia SV, Embleton ND, Rankin J. A systematic review of studies of quality of life in children and adults with selected congenital anomalies. Birth Defects Res A Clin Mol Teratol. 2012;94(7):511–20.
44. Wang Y, Hu J, Druschel CM, Kirby RS. Twenty-five-year survival of children with birth defects in New York state: a population-based study. Birth Defects Res A Clin Mol Teratol. 2011;91(12):995–1003.
45. Shin M, Kucik JE, Siffel C, Lu C, Shaw GM, Canfield MA, et al. Improved survival among children with spina bifida in the United States. J Pediatr. 2012;161(6):1132–7.
46. Brodwall K, Greve G, Leirgul E, Klungsøyr K, Holmstrøm H, Vollset SE, et al. The five-year survival of children with down syndrome in Norway 1994–2009 differed by associated congenital heart defects and extracardiac malformations. Acta Paediatr Int J Paediatr. 2018;107(5):845–53.
47. Cassina M, Ruol M, Pertile R, Midrio P, Piffer S, Vicenzi V, et al. Prevalence, characteristics, and survival of children with esophageal atresia: a 32-year population-based study including 1,417,724 consecutive newborns. Birth Defects Res A Clin Mol Teratol. 2016;106(7):542–8.
48. Glasson EJ, Jacques A, Wong K, Bourke J, Leonard H. Improved survival in down syndrome over the last 60 years and the impact of perinatal factors in recent decades. J Pediatr. 2016;169:214–20.e1.
49. Halliday J, Collins V, Riley M, Youssef D, Muggli E. Has prenatal screening influenced the prevalence of comorbidities associated with down syndrome and subsequent survival rates? Pediatrics. 2009;123(1):256–61.
50. Chua GT, Tung KTS, Wong ICK, Lum TYS, Wong WHS, Chow CB, et al. Mortality among children with down syndrome in Hong Kong: a population-based cohort study from birth. J Pediatr. 2020;218:138–45.
51. Meyer RE, Liu G, Gilboa SM, Ethen MK, Aylsworth AS, Powell CM, et al. Survival of children with trisomy13 and trisomy 18: a multi-state population-based study. Am J Med Genet A. 2016;170A(4):825–37.
52. European network of registries for the epidemiologic surveillance of congenital anomalies (EUROCAT). 2023. https://eu-rd-platform.jrc.ec.europa.eu/eurocat/eurocat-data/prenatal-screening-and-diagnosis_en.
53. Samadirad B, Khamnian Z, Hosseini MB, Dastgiri S. Congenital anomalies and termination of pregnancy in Iran. J Pregnancy. 2012;2012:574513.
54. European network of registries for the epidemiologic surveillance of congenital anomalies (EUROCAT). 2023. https://eu-rd-platform.jrc.ec.europa.eu/eurocat/eurocat-data/key-public-health-indicators_en.

Aetiology

3

Contents

The prevention of birth defects requires prior knowledge of the causal factors involved. In general, the aetiology of congenital anomalies is still largely unknown. Although much has generally been learnt about the experimental outcome of malformations in lab-based investigations, very little is known about causes in human.

Congenital anomalies can be caused by genetic background, environmental factors, or as an interaction between both genetic and environment. Genetic factors include chromosomal abnormalities, gene mutations, and inherited disorders whereas environmental factors include maternal infections, exposure to various teratogens, and poor prenatal and maternal health care during pregnancy [1–4]. The causes of congenital anomalies can, therefore, be divided into three major categories:

3.1 Genetic Factors

A great number of congenital anomalies with known causes have genetic basis. The genetic disorders are usually congenital, or they may have delayed presentations at later ages in the adulthood. The genetic disorders can be inherited from parents or may be a result of a new independent mutation in individuals.

Three main types of chromosomal abnormalities have been shown to be associated with congenital malformations including numerical chromosomal abnormalities (i.e. Turner syndrome, trisomy of autosomes, trisomy of the sex chromosomes), structural chromosome abnormalities (i.e. translocation, deletion), and anomalies caused by mutant genes and inherited disorders [3].

S. Dastgiri, *Congenital Anomalies in Iran*, https://doi.org/10.1007/978-981-97-6274-3_3

3.2 Environmental Factors

Although the human embryo is well protected in the uterus, certain environmental agents may cause developmental disruptions following maternal exposure to them. It is, however, known that some of human birth defects result from the disruptive actions of drugs, viruses, and other environmental factors. These factors can either indirectly affect the unborn child by causing disease in the mother, or alternatively, produce chromosomal damage or mutation in the developing foetus [3, 4].

Maternal exposure to alcohol consumption, marijuana and tobacco use, environmental chemicals and agents including mercury, anaesthetic gases, ionizing radiation, some viruses (i.e. Rubella, Cytomegalovirus, Herpes simplex, and Varicella), and therapeutic drugs are among the reported teratogens and environmental factors for birth defects. Some other predisposing factors associated with birth defects include poor care before and during pregnancy, advanced maternal age, and nutritional deficiencies of vitamins such as folic acid or high levels of vitamin A, maternal obesity, and suffering from diabetes mellitus in pregnancy [1, 2, 5, 6].

The list of the definite or probable teratogenic agents is always growing. Although this list is still growing, it accounts for only a small proportion of the causes of congenital anomalies.

3.3 Interrelated Genetic and Environmental Factors

Many common congenital anomalies are caused by genetic and environmental factors acting together. Unfortunately, the distinction between genetically caused and environmentally caused defects is blurred at times since environmental dangers may induce a mutation in the sperm or the egg that later causes a genetic disorder in the child. Multifactorial inheritance is the most common cause of congenital anomalies.

References

1. World Health Organization. Congenital anomalies fact sheet. 2023. https://www.who.int/news-room/fact-sheets/detail/birth-defects.
2. Centers for Disease Control and Prevention. Facts about birth defects. 2023. https://www.cdc.gov/ncbddd/birthdefects/facts.html#causes.
3. Tobias ES, Connor JM, Ferguson-Smith MA. Essential medical genetics. Wiley-Blackwell; 2011.
4. Moore KL, Persaud TVN, Torchia MG. The developing human. In: Clinically oriented embryology. Saunders; 2012. p. 167–200.
5. European network of registries for the epidemiologic surveillance of congenital anomalies (EUROCAT). https://eu-rd-platform.jrc.ec.europa.eu/eurocat/prevention-and-risk-factors_en.
6. Carmichael SL, Shaw GM. Maternal corticosteroid use and risk of selected congenital anomalies. Birth Defects Res. 2018;110(14):1017–27.

Principles of Control and Prevention

4

Contents

Birth defects create a significant burden on affected individuals, their families, and substantial economic and emotional burden on the society. The financial cost of treating children (mainly surgical correction) with birth defects, and the tragic effects and disruption of normal family life associated with deaths caused by congenital anomalies are also enormous for people in both developed and less developed countries. Therefore, preventive measures, early diagnosis, effective interventional programmes, appropriate clinical and care management are essential to control and prevention the occurrence, reduce the burden and impact of these disorders at the community level.

Although the causes of congenital abnormalities are largely unknown, preventive measures are available for about 70% of congenital abnormalities [1, 2]. Preventive interventions of birth defects should therefore be set as a priority for the public health system of the society.

A wide range of interventional strategies and approaches for the control of congenital and genetic disorders in the Eastern Mediterranean Region of the World Health Organization have been recommended by the experts group of specialists in the field of community genetics (Table 4.1) [3].

The major strategies for efficient control and prevention of congenital anomalies include preconception counselling, prenatal screening, maternal health care, regulations to reduce exposure to teratogenic substances in the workplace and environment, and public health genetic knowledge and education. The details of these

S. Dastgiri, *Congenital Anomalies in Iran*, https://doi.org/10.1007/978-981-97-6274-3_4

Table 4.1 Interventions recommended for the control of the congenital and genetic disorders [3]

Interventions
Interventions during the childbearing and preconception (including premarital) period
• Immunization for Rubella and Hepatitis B virus
• Fortification with vitamin B12, folic acid, iodine and iron
• Screening for carriers of common autosomal recessive disorders including beta thalassemia and sickle cell anaemia
• Assessing maternal age distribution, genetic family history, and referral of high-risk families
• Screening and treatment of infections such as syphilis and HIV/AIDS
• Screening, diagnosis and management of diabetes, anaemia and hypertension
• Avoidance of tobacco use
• Increasing public awareness of risk factors, such as unhealthy diet, sedentary lifestyle and common genetic risk factors
Interventions during pregnancy
• Early detection and management of maternal conditions such as diabetes
• Early detection and management of infections
• Avoidance of teratogens (infections such as toxoplasmosis, drugs)
• Prenatal screening by maternal serum markers in first trimester and by ultrasonography
• Prenatal diagnosis with/without termination of pregnancy
• Care of fetus for conditions such as Rh incompatibility
• Avoidance of tobacco use and exposure to pollution
• Supplementation with iron and folate
Interventions after birth
• Newborn biochemical screening for congenital hypothyroidism, phenylketonuria (PKU), galactosaemia, sickle cell disorder, glucose-6-phosphate dehydrogenase (G6PD) deficiency, congenital adrenal hyperplasia, methyl coenzyme dehydrogenase deficiency
• Newborn screening for hearing impairment, congenital hip dislocation, isolated cleft palate, heart defects and other clinically identifiable congenital disorders
• Stillborn examination and investigations
• General newborn management of identified conditions, and paediatric surgery to correct malformations
• Rehabilitation and palliative care
• Family support, including bereavement
• Diagnosis of congenital and genetic disorders, with management and counselling families on future reproductive options
• Extended family screening testing and counselling

strategies are often classified in the following levels of prevention:

4.1 Primordial Prevention

The primordial level of prevention is defined as the "conditions, actions, and measures that minimize hazards to health and that hence inhibit the emergence and establishment of processes and factors (environmental, economic, social, behavioural, cultural) known to increase the risk of disease" [4]. Primordial methods are usually acting on healthy persons avoiding the birth defects arising in the first place by reducing the susceptibility or decreasing the exposure to risk factors that exist in the environment, work, society, culture, etc.

4.2 Primary Prevention

This level of prevention is defined as the "aims to reduce the incidence of disease by personal and communal efforts, such as decreasing environmental risks, enhancing nutritional status, immunizing against communicable diseases, or improving water supplies" [4].

The primary prevention approaches have to be taken preconceptionally. They are cheaper and the most effective plans in practice. Using these methods, congenital anomalies can be prevented by family planning to avoid extreme maternal age, by immunization against maternal infectious diseases that may cause birth defects, by genetic counselling to parents who have had an affected child, by reducing the risk of environmental factors (i.e. limitation of non-urgent abdominal X-ray in women of child-bearing age), and by controlling and supervising of therapeutic and non-therapeutic drug sales to pregnant women [5–7].

Success in primary prevention is recently evidenced by some epidemiological observations. Accordingly, there appears to be an association of congenital anomalies, particularly neural tube defects, to diet and the role of vitamin deficiency. Some studies have specifically shown that multivitamin supplements containing folic acid, before and during early pregnancy, may reduce the risk of neural tube defects [6, 7].

4.3 Secondary Prevention

It is defined as the "measures available to individuals and communities for the early detection and prompt intervention to control disease", "it aims to reduce the prevalence of disease by shortening its duration" [4]. Secondary prevention is acting on ill persons. Procedures in this level are quite efficient, but expensive and sometimes unacceptable by some cultures. They are mainly used over the prenatal period.

Prenatal diagnosis may be performed using biochemical and genetic analysis of foetal cells taken from the amniotic fluid, chorionic villi biopsy material, maternal peripheral blood, or even by direct inspection of the foetus by some specific techniques. However, the most common and popular method used for prenatal diagnosis is ultrasonography. This method can easily be accessible and is a non-invasive technique. The only limitation of this technique is that in most cases diagnosis is not feasible before the third trimester.

Secondary prevention approaches are taken after prenatal diagnosis of a congenital anomaly to avoid birth of affected babies. These approaches include medical treatment, intrauterine surgery for some types of anomalies (i.e. hydrocephaly), or therapeutic termination of the affected pregnancy where there has been proven exposure to harmful factors in early pregnancy or where there is a high risk of hereditary abnormality [8].

4.4 Tertiary Prevention

The methods in tertiary prevention consist of the "measures aimed at softening the impact of long-term disease and disability by eliminating or reducing impairment, disability, and handicap; minimizing suffering; and maximizing potential years or useful life" [4]. Tertiary prevention methods are postnatally acting on affected persons to avoid complications of the defects. In this level of prevention, many of the measures (i.e. rehabilitation techniques for some groups of congenital anomalies) are insufficient and expensive. They are, however, the only way to improve the quality of life of children born with congenital anomaly. Thus, the limited available resources of tertiary prevention must be rationally distributed and used in the community.

4.5 Quaternary Prevention

Quaternary level of prevention is defined as "procedures and policies that identify individuals and groups at risk of over-diagnosis or overmedication, and that decrease excessive medical and sanitary intervention" [4].

Clinical management and healthcare support for many types of congenital anomalies are usually expensive in every society. Quaternary prevention for birth defects, therefore, includes activities and actions taken to identify families and patient with an anomaly at risk of over-medicalization to mitigate or avoid the excessive intervention or from new unnecessary medical invasion.

***

In all the above levels of prevention programmes, genetic knowledge and education must be essentially considered for improving the public understanding of the causes and prevention of congenital anomalies, and to ensure that affected individuals and families receive optimal care and support.

All control programmes for congenital anomalies should interact and operate across various levels of prevention because the natural history of birth defects indicates there are largely conceptual and practical overlapping issues and procedures among different levels for the control and prevention of these disorders.

***

Nowadays, advances in medical procedures and genetics technology have provided new opportunities for screening, diagnosis, and treatment of congenital anomalies. Prenatal diagnosis of birth defects can be performed using various screening and diagnostic tests including ultrasound, amniocentesis, chorionic villus sampling, and some non-invasive prenatal testing. Postnatal diagnosis involves physical examination, imaging technologies and some genetic procedures [9–11]. Developed countries have remarkable achievements and experiences on these procedures and technologies for the control and prevention of birth defects. This may help public health authorities, health service planners, and non-government organizations (including patient/parent support groups) in developing countries for those challenges needed in working on the control and prevention of congenital anomalies in these areas.

References

1. Czeizel AE. Prevention of congenital anomalies by periconceptional multivitamin supplementation. Br Med J. 1993;306:1645–8.
2. Czeizel AE, Intody Z, Modell B. What proportion of congenital anomalies can be prevented? Br Med J. 1993;306:499–503.
3. World Health Organization. Expert meeting on the prevention of congenital and genetic disorders in the Eastern Mediterranean Region, London, United Kingdom of Great Britain and Northern Ireland 29–31 July 2016. http://applications.emro.who.int/docs/IC_Meet_Rep_2016_EN_18989.pdf?ua=1.
4. Porta M. A dictionary of epidemiology. edited for the International Epidemiological Association by Miquel Porta; associate editors, Sander Greenland, Miguel Hernán, Isabel dos Santos Silva, John M. Last; assistant editor, Andrea Burón. Oxford University Press; 2014.
5. European network of registries for the epidemiologic surveillance of congenital anomalies (EUROCAT). https://eu-rd-platform.jrc.ec.europa.eu/eurocat/prevention-and-risk-factors/primary-prevention_en.
6. Centers for Disease Control and Prevention. Facts about birth defects. 2023. https://www.cdc.gov/ncbddd/birthdefects/facts.html#Prevention.
7. World Health Organization. Congenital anomalies fact sheet. 2023. https://www.who.int/news-room/fact-sheets/detail/birth-defects.
8. Center for Disease Control and Prevention. Facts about birth defects. 2023. https://www.cdc.gov/ncbddd/birthdefects/types.html.
9. American College of Obstetricians and Gynecologists. Practice bulletin no. 163: screening for fetal aneuploidy. Obstet Gynecol. 2016;127(5):e123–37.
10. World Health Organization. WHO recommendations on antenatal care for a positive pregnancy experience. 2023. https://www.who.int/publications/i/item/9789241549912.
11. World Health Organization. Community genetics services: report of a WHO consultation on community genetics. Geneva: World Health Organization; 2011.

Detection of Congenital Anomalies 5

Contents

For the detection, and for a true evaluation of an anomaly in a child, a comprehensive history and physical examination followed by further examinations and tests in the later ages are needed. Since the risk of having a syndrome increases with the number of defects, a full familial history of congenital anomalies, history of miscarriages, stillbirths and spontaneous abortion, and a comprehensive consultation by a genetic specialist is typically suggested for any infant having a single anomaly or a combination of multiple defects.

5.1 Familial Aggregation

Genetic epidemiology methods routinely examine the impact of genetic factors, environmental exposures, and the interaction in between these two elements on the occurrence of a disease or a trait, and that how and why diseases, traits, and the relevant causes cluster in certain groups.

Familial aggregation, as part of the genetic epidemiology, is to investigate a group of individuals with a specific defect for identifying whether rela-

S. Dastgiri, *Congenital Anomalies in Iran*, https://doi.org/10.1007/978-981-97-6274-3_5

tives have an excess occurrence risk of the same anomaly. In summary, familial aggregation and susceptibility would indicate whether a certain anomaly or disease runs in families/generations.

Family history information is an important risk assessment tool and useful in assessing the familial aggregation. It is a risk tool for some groups of birth defects showing the impact of shared genetic variation, shared exposures to environmental factors, or shared life style and behaviours. Obviously, the blood relatives share a huge part of their genes with one another compared to unrelated individuals in the population; therefore, we see some defects clustered in some families. However, familial aggregation does not necessarily mean there must be a genetic origin for those congenital anomalies clustered in families. The clusters might occur as a result of a shared environmental exposure in the family, or even by chance alone. Thus, whatever the underlying predisposition is in a cluster of multifactorial condition in families, there is an opportunity to identify and pull out the individuals in the family who are at higher risk for a certain defect.

The prenatal history using particular tests may uncover some specific exposures and certain aetiologic factors. The history taken should include medical and obstetric history including the duration of gestation, history of stillbirths and miscarriages, prenatal care, and maternal exposures (i.e. alcohol, prescribed or illicit drugs, cigarettes, fevers, illnesses, chemicals, radiation). The family should be asked about consanguinity marriage (for the possibility of autosomal-recessive disorders), age of the older parents (for the possibility of chromosome aneuploidies and autosomal-dominant disorders, etc.). When, therefore, the familial aggregation shows an increased risk, genetic testing may be used as an opportunity to identify high-risk individuals for further preventive interventions.

5.2 Surveillance of Congenital Anomalies

As defined earlier, surveillance is a public health programme that collects, monitors, analyses, interprets, and disseminates data systematically in a timely manner and that allows for planning, implementation, and evaluation of health strategies.

Surveillance of congenital anomalies is a critical part of healthcare management of birth defects. The programme aims to reduce the burden of these disorders in individuals, families, and population. Accordingly, the surveillance activities examine the data on the occurrence and characteristics of congenital anomalies to explore the trends of occurrence, to evaluate the impact of the interventional and preventive efforts, and to provide essential information for public health policymakers [1].

Birth defects surveillance programmes might typically be either population-based registry or a hospital-based monitoring system, and can be an active or passive case ascertainment approach, or a hybrid of the two systems that collect data on infants born with congenital anomalies or foetal anomalies. The surveillance programmes typically rely on data reporting by healthcare providers, clinicians, laboratories, vital records, discharge data, insurance databases, and other sources.

The geographical area for the population-based registries might be a city, a region, or the whole population of a country. In these programmes, the identified birth defects occurring within the defined geographical area are registered regardless of delivery site (i.e. hospitals or maternity facilities and home) within a defined time period.

Hospital-based surveillance programmes are typically used for monitoring some specific groups of congenital anomalies that occur in selected facilities and hospitals in a defined geographic area (i.e. a state, province, or county).

In any surveillance programmes, the case ascertainment approaches can be either active, passive, or hybrid methods. In an active case ascertainment, the registry trained personnel are hired for the collection, management, and abstraction of the data whereas in the passive case ascertainment method, hospital personnel once identify a foetus or neonate with a defect reports this information to the registry. The anomaly cases can even be ascertained by linking administrative databases in the hospitals or maternity facilities in the passive method.

Hybrid case ascertainment approaches are usually used as the combination of passive case ascertainment of most types of birth defects, with active case ascertainment of some specific predefined anomalies. Hybrid methods may also be used as a quality control tool for the verification of the data validity [2].

5.3 The Key Components of a Registry of Congenital Anomalies

Typically, a registry of congenital anomalies is formed based on the following core features:

- What is the programme model: hospital-based or population-based?
- What are the case ascertainment methods: active, passive, or hybrid?
- Which organizations will be participating?
- Estimating the required annual budget?
- Who should be included in the executive committee?
- Who should be included in the scientific committee?
- How many expert personnel are needed? level and field of education?
- Description of organizational duties and responsibilities of each expert personnel?
- Definition and codes of anomalies subject for inclusion in the registry?
- The criteria for diagnostic validity?
- List of the congenital anomalies for inclusion in the registry?
- How are congenital anomalies coded and recorded?
- Minimum variables and data required for recording? any additional variables?
- How will be the validity of the data evaluated?
- Data collection and management procedures?
- Data collection tools? paper, electronic, online?
- Statistical methods required for the analysis of the data?
- Dissemination and reporting protocols?
- Medical ethics considerations and confidentiality matters?
- Data access and availability?

5.4 Challenges in Congenital Anomaly Surveillance Systems

Despite the importance of surveillance systems in the control of congenital anomalies, there are some challenges in the systematic collection and administration of surveillance programmes. Some of the challenges include: inconsistencies in the case definition and classification, underreporting and incomplete data, limited data on potential risk factors of congenital anomalies, and limited resources for personnel, infrastructure, and funding. These major challenges make it difficult to compare data in surveillance systems across the countries and regions.

Continued efforts to improve the surveillance of congenital anomalies is always an essential core part for reducing the burden of these disorders in individuals, families, and in the whole population. There are, however, some solutions that can be pursued to tackle those challenges and enhance the monitoring of congenital anomalies. These include:

- Standardizing the classification and case definitions would enhance the quality and comparability of data among surveillance systems.
- The integration of data from various sources including medical records, genetic testing, and birth certificates has the potential to improve the completeness and accuracy of the data.
- The inclusion of data on potential risk factors (i.e. maternal medication history and environmental exposures) may improve our understanding of the aetiology of birth defects.
- New technologies (i.e. electronic health records and machine learning) for data collection and management could improve the effi-

ciency and accuracy of the surveillance programme.
- Collaboration and data sharing among different surveillance systems are useful to identify the effectiveness of prevention strategies in the community by examining the patterns and trends of the occurrence of congenital anomalies at the population level.
- Evaluation procedures are essential part of any programme to monitor effects of preventive services implemented for congenital anomalies in order to identify and correct shortcomings and to assure the quality of the data. This will provide essential feedback required to improve the weaknesses. Periodic evaluation will have to, therefore, take place in each surveillance programme on a regular basis [2].

5.5 Surveillance Programmes and Registries

There are a number of scientific societies and associations that working on the aetiology, biological processes, pattern of occurrence, and prevention and clinical aspects of congenital anomalies. There are, however, a limited number of entities that registering the data on the occurrence of congenital anomalies in countries and regions. Most of them working nationally or in a local level inside the countries, and some working exclusively on specific groups of defects (i.e. Down syndrome, Spina Bifida). The following organizations are of main institutions working as the registries of congenital anomalies continentally or internationally:

5.6 International Clearinghouse for Birth Defects Surveillance Research (ICBDSR)

This is a voluntary and non-governmental consortium of 45 population- and hospital-based surveillance programmes across the globe (including Europe, North, Central and South America, Asia, New Zealand, and Australia) for investigating the occurrence of at least one of 39 various congenital anomalies, and collaborating in the prevention programmes, possible causes and adverse outcomes related to birth defects. ICBDSR was originally established in 1974 to facilitate the exchange of data on birth defects among the representatives of the member countries and institutions. The programme has been affiliated with the World Health Organization (WHO) since 1986. ICBDSR publishes annual reports for the occurrence of congenital anomalies based on the data provided regularly by the member institutions and countries. Some collaborative research programmes are also carried out using the ICBDSR registries for publication in international journals. There is an annual meeting held every year in one of the member countries to discuss the scientific and business matters by the registry directors. There are workshops run by the scientists from ICBDSR to provide expert consultation and assistance for existing registries to investigate outbreaks and for the establishment of new monitoring systems. The ICBDSR headquarters is located in Rome, Italy, for coordinating the production of reports, newsletters, and monitoring activities and collaborative studies. An initial list of congenital anomalies for monitoring based on the ICD-10 codes is recommended by ICBDSR. This list is presented in the Appendix B.2. The details of the history, size of the population coverage, legislation, funding, directors, and case ascertainment methods can be found for each programme on the ICBDSR website [3].

5.7 European Network of Registries for the Epidemiological Surveillance of Congenital Anomalies (EUROCAT)

This is a programme supported by the European Union, represented by the Commission of the European Community for the epidemiological surveillance of congenital anomalies in Europe. High-quality multiple sources are implemented in the EUROCAT registries. Terminations of pregnancy as well as births are ascertained in the EUROCAT network. The main objectives are to detect and investigate trends in the frequency of congenital anomalies that could be due to environmental teratogens or mutagens and to evaluate the effectiveness and efficiency of neonatal and perinatal health services. The specific objectives of the EUROCAT registration are:

- To provide baseline epidemiologic information on congenital anomalies in Europe
- To detect and investigate trends in the frequency of congenital anomalies in order to assess the impact of known or suspected risk factors particularly related to exogenic agents (i.e. drugs, environmental and occupational exposure)
- To evaluate the effectiveness and efficiency of health services (i.e. primary prevention, prenatal diagnosis and termination of pregnancy, treatment)
- To provide a well-documented database for aetiologic and clinical research
- To act as information centre which could respond to specific needs, such as the assessment of the impact of environmental accidents or change, or the suspicion of teratogenic influences from food, drugs, or other exposures
- To act as a catalyst for setting up of registries throughout Europe collecting comparable and standardized data

EUROCAT holds regular meetings and workshops by member representatives to discuss various issues about congenital anomalies. There are also various studies and investigations carried out by EUROCAT scientists in the area of aetiology, prevention, treatment, and care of affected individuals and families. Prevalence data of congenital anomalies are freely available from the member states and countries on the EUROCAT website. The groups of congenital anomalies reported by the EUROCAT network for prevalence data are listed in the Appendix B.3 [4].

5.8 Latin American Collaborative Study of Congenital Malformations (ECLAMC)

This is a clinical and epidemiological investigation programme on the occurrence and causes of congenital anomalies in the Latin American hospital births established in 1967. The ECLAMC collects data and publishes information on the occurrence of congenital anomalies from Latin American countries [5].

References

1. European network of registries for the epidemiologic surveillance of congenital anomalies (EUROCAT). 2023. https://eu-rd-platform.jrc.ec.europa.eu/eurocat/data-collection_en.
2. Birth defects surveillance: a manual for programme managers, second edition. Geneva: World Health Organization; 2020.
3. International Clearinghouse for Birth Defects Surveillance Research (ICBDSR). 2023. http://www.icbdsr.org.
4. European Network of Population-Based Registries for the Epidemiological Surveillance of Congenital Anomalies (EUROCAT). 2023. https://eu-rd-platform.jrc.ec.europa.eu/eurocat.
5. Latin American Collaborative Study of Congenital Malformations (ECLAMC). 2023. http://www.eclamc.org.

Part II

6 Background

Content

The aim of this section of the book is to document the long-term running of a surveillance system within a developing country, Iran, by describing the experience of Tabriz Registry of Congenital Anomalies (TRoCA) followed by a brief description of the Epidemiology and Surveillance of Congenital Anomalies in Iran (ESCAI) and Tabriz Foundation for Public Health Genetics (TFPHG).

The report describes the processes in setting up the registry and utilization. It will attempt to draw practical lessons of relevance to all those involved in birth defects surveillance. Legislation, funding, data management and reporting, confidentiality issues, and evaluation of TRoCA are discussed in this section.

This report should prove useful to both practitioners and policy makers of birth defects control programmes. The report is also intended to introduce the key components for the development, implementation, and improvement of a surveillance programme for congenital anomalies. This might, particularly, be useful for those willing to establish a similar registry in the countries with limited resources.

6.1 Report Frame

The information gathered for this report was obtained in a systematic way. In gathering information, the focus points included to describe the processes involved in setting up of the TRoCA, the processes involved in the running of the TRoCA, the changes made to the programme over the 23 years, and the strengths and weaknesses of the programme.

To do this, all written materials relating to the TRoCA was reviewed. This included publications, presentations, correspondence, lectures, interviews, meeting notes, and prevention activities throughout the TRoCA programme from the beginning. Two meetings of the steering committee were held every year discussing the activities of the TRoCA programme in the area. The minutes of those meetings was reviewed. Key

S. Dastgiri, *Congenital Anomalies in Iran*, https://doi.org/10.1007/978-981-97-6274-3_6

individuals and end users of TRoCA were also interviewed about activities of the programme. The following main points are particularly been tried to address using the data collected in TRoCA registry over the past 23 years:

- To provide an overview of the epidemiology of congenital anomalies in the area
- To describe the natural history of congenital anomalies
- To determine the impact of foetal, maternal, and environmental variables on the natural history and survival pattern of congenital anomalies
- To introduce a model for prevention of congenital anomalies in the area

7 Tabriz Registry of Congenital Anomalies

Contents

7.1 How the Registry Was Initiated?

Until the recent years, there were no reliable data available about the prevalence, aetiology, and preventive strategies of congenital anomalies in Iran. Beginning from 2000, a research project on the epidemiology of congenital anomalies was carried out aiming to document the epidemiologic features of congenital anomalies in the northwest region of Iran, and to provide the baseline components to set up a regional registry of birth defects. Accordingly, Tabriz Registry of Congenital Anomalies (TRoCA) was then established. TRoCA is the first programme for the surveillance of birth defects in the country. It has since been running for more than two decades now.

Initially in 2000, some of the registry systems including the European network countries (EUROCAT), United States and United Arab Emirates, and International Clearinghouse for Birth Defects Surveillance and Research (ICBDSR) were studied in terms of data collection, process, analysis, and evaluation of the systems to determine the minimum requirements for setting up a local registry in Iran. Accordingly, it was agreed that the type of anomalies, basic demographic information, history of pregnancy, birth data, family history, laboratory data, and

S. Dastgiri, *Congenital Anomalies in Iran*, https://doi.org/10.1007/978-981-97-6274-3_7

possibly exposure information in the family is preferred to be included in TRoCA registry.

TRoCA uses a "passive" method of data collection. The end users preferred that the responsible persons (as registrar) for collection be nurses, midwives, or medical documentation officers of the local hospitals or health centres.

The principal aims of TRoCA programme are to facilitate the collection of essential information for documenting the burden of birth defects, as well as the detection of any unexpected problem related to the occurrence of congenital anomalies in the area that may require intervention and follow-up. TRoCA examines the occurrence and patterns of birth defects and looks for possible local causes, influencing factors and aetiology in the population, and creates information essential for genetic consultation, prognostic values, and planning and evaluating the prevention measures, particularly in high-risk groups. The programme ultimately trying to provide further insight into the medical and healthcare needs of children with anomaly and families, and potential for prevention and control measures. Purposes of the TRoCA, in summary, are to:

- Register the occurrence, rates and patterns of selected birth defects in the region
- Identify, if any, causative and influencing factors associated with the rate and pattern of congenital anomalies
- Prepare epidemiological indexes to indicate the magnitude and trends over time
- Monitor emerging or unusually high occurrences of congenital anomalies in the area
- Identify further possible interventions for the control and preventing birth defects in the region based on the findings
- Make valid data available to policy makers and researchers
- Plan and implement preventive and control strategies to prevent selected anomalies
- Evaluate prevention and control strategies
- Provide baseline framework to set up a biobank for congenital anomalies in the region
- Provide baseline framework and information to set up the national population-based surveillance of congenital anomalies

As indicated above, the TRoCA programme examines the rates and patterns of various types of congenital anomalies and looks for possible local causes and influencing factors in infants diagnosed in maternity hospitals, children's hospitals, or genetic clinics. Accordingly, TRoCA aims to introduce and implement some prevention programmes for the control of congenital anomalies for families who have a history of an anomaly in the family, educational programmes for youth, and preconceptional educational programmes for young couples. To achieve these goals, TRoCA focuses mainly on the following approaches:

- The use of genetic knowledge for prevention (i.e. the nature of genetic diseases, modes of inheritance, their prevention, and the current available approaches for management of a congenital anomaly and genetic disorder).
- Introducing and improving the genetic counselling and tests: many genetic tests are currently available in the area for those who need. However, there is a gap between the diagnosis and management of congenital anomalies and genetic diseases by individuals and families. For instance, according to the current policies in the country, termination of the pregnancy is only permitted for the life-threatening anomalies whereas many young couples willing and require termination of pregnancy if they are diagnosed with an even minor anomaly. The framework of counselling in TRoCA is to facilitate this demand and remove such barriers by considering the policies, social values, and beliefs in the population.

Under the TRoCA activities, a new Non-Governmental Organization (NGO) has been established in the area since 2013 aiming for the prevention (mostly the primary and tertiary levels of the preventive strategies) of selected groups of congenital anomalies and disabilities in the region. It is now called Tabriz Foundation for Public Health Genetics (TFPHG). More details will be illustrated in the next chapter.

7.2 Location, Organizational Capacity, Legislation, and Funding

The programme area is located in the northwest of Iran with a cool and dry climate that covers an area of approximately 47,830 km^2. The area population was estimated at 4,854,201 in 2021. The region has common borders with the Republics of Azerbaijan, Armenia, and Nakhchivan. The centre of the area is the city of Tabriz, one of the three major cities in Iran. Due to the role of this city as the capital in the area, individuals from various ethnic groups and religions live there. The most common occupations in the region are subsistence farming, small-scale trading, and some large industries. Most people live (nearly 70%) in the cities [1].

The public health system comprises a network of health centres and hospitals providing health care and medical services for the regional population. Health and medical facilities are all working under the Tabriz University of Medical Sciences. This is one of the five top universities in the country providing medical and health services for the population in the northwest of Iran. It covers 18 public university hospitals. There are also 9 private hospitals in the area collaborating with the university clinical staff and medical settings. TRoCA is run and administered under the rules and regulation of Tabriz University of Medical Sciences.

TROCA is located in Alzahra and Koodakan university hospitals of Tabriz University of Medical Sciences. Both are major university hospitals of Tabriz University of Medical Sciences and includes academic departments of obstetrics, gynaecology, paediatrics and child health (including units for growth and development, immunodeficiency, asthma and allergy, paediatric nephrology, paediatric nephrology, paediatric gastroenterology and hepatology, paediatric urology, paediatric cardiology, paediatric infectious diseases, paediatric surgery, paediatric pulmonology), neonatology, genetic diagnostic lab, radiology and imaging, pathology, and an NGO for breast cancer screening in the region (located in Alzahra hospital).

The TRoCA programme has been financially supported by the National Public Health Management Centre (a WHO collaborating centre in Iran), Tabriz University of Medical Sciences, Iran Ministry of Health and Medical Education, and some local and regional Non-Governmental Organizations.

7.3 Data Collection, Management, and Dissemination

The TRoCA programme covers live births, stillbirths, and therapeutic termination of pregnancies in the area diagnosed with one of the anomalies in this population. All infants are routinely examined by a gynaecologist, obstetrician, neonatologist, or paediatrician at birth, and at the time of hospital discharge for possible intervention and treatment. The examinations include assessment of general health, maturity, and congenital anomalies. Medical records at delivery and routine hospital discharge forms are the main data sources from which the registrar at the facilities collects the data. The definition of the congenital anomalies for the purposes of this programme based on the standard coding system of the International Classification of Diseases (ICD) and the British Paediatric Association under one of the following main headings according to the primary diagnosis of anomaly (Appendix A.1) [2]:

- Nervous system anomalies
- Genitourinary tract and kidney
- Anomalies of limb
- Chromosomal anomalies
- Cleft lip with/without palate
- Congenital heart disease
- Musculoskeletal and connective tissue anomalies
- Digestive system anomalies
- Eye and ear anomalies
- Other anomalies

If needed, a pathologic confirmation of the defect is used as criteria of inclusion of foetal deaths or stillbirths in this registry.

Five full-time nurses, midwives, and medical officers work closely with the TRoCA director in the management of the programme including:

- Planning and running the meeting and publicity to involve all relevant staff in the clinics, hospitals, and other entities and centres.
- Collection and filling the forms.
- Ensuring the completeness and accuracy of data collection and ensuring that all eligible people are entered into the programme. This process will require regular visits to all relevant hospitals and clinics to ensure that all cases of congenital anomalies in the study population are included.
- Coding the data and entering into the database.
- Analysis of the data.
- Coordinating the interventional programmes for high-risk individuals and families.

For inclusion in the registry, an infant must have been born to a woman who resided in the defined population area at the time of the infant's birth.

A variety of epidemiological and statistical methods are used to analysis the data at TRoCA. Total prevalence (with 95% confidence intervals) is routinely calculated by dividing the numerator (registered cases of congenital anomalies) by the relevant denominator (total live and stillbirths) for the same period of time at the same place. An infant/foetus with more than one anomaly is counted once only in the numerator. For other purposes, more analysis may also be calculated including time trend analysis and estimates for the relative frequencies. Kaplan-Meier estimates (with 95% confidence intervals) are calculated to provide a summary of the survival experience of children with congenital anomalies.

The TRoCA programme has recognized that collecting the data by one specialist group can result in more consistent information than that by different specialist groups. Medical coders, for instance, with skills of abstracting data seemed to be the most competent individuals for collecting data in the registry.

The end users and hospital administration and authorities agreed to establish a database in an electronic layout is used to take the maximum advantage of collected data for intended purposes including the re-using of the data in various occasions.

In the registry, the programme officer abstracts the data and completes the standard form of the registry. Completed paper-based forms are sent to the centre of the registry. Transfer of data are done by a dedicated person allocated for the purpose. Registrar enters the data received from different sources into the computerized database designed for the purpose of the registry.

In the TRoCA, a form is designed based on its data elements and the list of variables defined in the registry. A user's manual for the registry system is available illustrating the procedures for using the registry data along with the mechanisms of access to the data and the list of authorized people for access.

The required data and information by the ICBDSR are regularly submitted to the headquarter using an online system dedicated by the ICBDSR. The list of congenital anomalies requested for inclusion in the ICBDSR database can be found in Appendix A.4.

In addition to the minimum data submitted annually to the ICBDSR, the TRoCA programme gathers data on the various types of anomalies as diagnosed, gender, date of diagnosis, mother age at delivery, family history of anomaly, familial marriage, date of death (if any), and exposure information from parents of all defected infants. Other women giving normal and healthy births in the facilities covered by the registry routinely complete similar data form. They are considered as matched control group when needed for research and other purposes.

There is no plan for regular dissemination of results to external audiences. However, findings have been reported in peer-reviewed journals, books (in Persian), and many lecture presentations in domestic and international meetings. Several student projects and thesis were carried out based on the TRoCA data and programmes. We have developed some preventive initiatives and programmes in the region. This includes the use of genetic knowledge for prevention purposes and improving the genetic counselling and tests in the community level. These were funded by the Regional Welfare Organization and performed through community campaigns in various cities, districts, and occasions in the region. Anecdotal reports suggest the public programmes and campaigns were well received.

7.4 Confidentiality of Data and Ethics

The TRoCA was approved by the Ethics Committee of the Tabriz University of Medical Sciences when the programme began in the year 2000. All data are handled accordingly. Personal information is anonymized, and principles of privacy and confidentiality are strictly considered in every part of the registration and data management. Strict safeguards are observed to prevent unauthorized access to the registry data particularly to the sensitive data including identity-related information.

7.5 Evaluation of the Registry

Evaluation procedures are essential part of the TRoCA programme to monitor effects of preventive services for congenital anomalies in order to identify and correct weaknesses. The programme can monitor the number of new cases in the coverage area using the registry of congenital anomalies. Patients can then be followed up to assess the extent to which prevention interventions in the population have been effective. Using the registry data, prenatal diagnosis services can also be assessed by birth prevalence rate of affected individuals, the choices that couples (at risk) make with respect to using prenatal diagnosis, the rate of prenatal diagnosis performed, the rate of births of children with avoidable conditions, and the rate of medical termination of pregnancies due to fatal defects. In addition, similar to any surveillance system, TRoCA applies the following attributes of a standard monitoring system for evaluating the registry [3]:

- Simplicity: the data collection forms are simple and easy to complete in TRoCA. Basic analysis can be performed relatively easily. Data coding and entering is, however, relatively complicated because of the large number of codes for the data gathered.
- Flexibility: TRoCA has been easily changed and adapted in the region after the original design was transferred from the ICBDSR to this regional programme in Iran.
- Acceptability: health system authorities, individuals and the public have accepted our programme. They have been supportive of TRoCA by funding us and showing their willingness for collaboration on various occasions.
- Reliability: high capture rates were not consistently maintained because of the difficulties in diagnosis methods, procedures, and financial problems (mainly in the recent few years).
- Utility: the cost of collecting, entering, and processing the data have been almost affordable. However, in the most recent years we are having some difficulties in the aids and funds we received.
- Sustainability: the lack of funding for dedicated staff was not sustainable. Ongoing data collection for updating the surveillance system has been possible and available. There is no formal mechanism for obtaining feedback from data users to improve the system. The lack of formal evaluation system resulted in continued collection of poor data quality in some occasions. Unplanned changes (i.e. changes in personnel working for TRoCA) have resulted in some problems in the collection of data in some occasions.
- Timeliness: basic analysis can be carried out to produce up-to-date information in a couple of weeks after the data collected.
- Sensitivity: TRoCA is able to detect epidemics of congenital anomalies in its defined population.

- Representativeness: our system is able to describe the occurrence of birth defects over time and its distribution in the community by various influencing factors (i.e. place, time, and person) [4].

Although there is no formal plan for the evaluation of the registry, a minimum of two meetings of the "registry committee" are held every year discussing the activities of the programme in the area. Key individuals and end users of the programme are also identified and interviewed about the activities, shortcomings, and limitations of the programme along with these meetings together with issues relating the interpretation, reporting, and use of the data.

7.6 How Did TRoCA Perform?

The TRoCA programme has successfully achieved its main objective [5]. This includes: the detection of epidemics (if any) of congenital anomalies in the population covered, time trends analysis of the occurrence of birth defects over two decades, estimates of the prevalence for selected groups of anomalies by individual and geographical variations, aetiological investigations, implementation of preventive and genetic services to families who have a history of an anomaly or defect in the family, and preconceptional programmes for young couples. The prevention programmes are administered under the Tabriz Foundation for Public Health Genetics which will be described later in this chapter.

Beside those achievements, the TRoCA porogramme has, however, had some inevitable limitations and weaknesses in some parts of its activities took place over years:

- The very small frequency and the rare nature of some groups of anomalies influence the epidemiological power of the rates and occurrence patterns of various types of congenital anomalies provided by the programme.
- Some anomalies are routinely not diagnosed until some time after birth. For this reason, ascertainment of those defects is inevitably delayed.
- All the registries (including TRoCA) that are based on the hospital data gathered during the medical and clinical management of the subjects are likely to be incomplete.
- The improvement of the diagnosis and identification of birth defects over time have resulted in the variable validity of data recorded by TRoCA.
- Setting up a bio-bank is one of the very costly objectives of the TRoCA programme. There is no plan yet, because of financial limitations, to establish this biological database for the birth defect cases in the region registered at TRoCA.

7.7 Major Challenges

Despite the achievements and successes of the TRoCA programme, there are challenges faced by the registry in its activities for an effective surveillance and monitoring of congenital anomalies in the population under coverage [5].

TRoCA faces the challenges mainly because of the lack of full-time personnel. The programme relied on staff for whom the TRoCA was not a core part of their responsibilities. Lack of full-time staff for data collection and reliance on busy staff at the facilities of the registry resulted in periods of poor data capture. The lack of funded individuals to perform the quality data analysis might also result in under-use of the data in TRoCA programme. Ideally, a professional data analyst would provide regular reports to the local and wider public health community and will support parties willing to use the TRoCA data for prevention, clinical, and administrative purposes.

Allocating fully funded individuals (preferably from nurses, midwives, or medical documentation officers) for the TRoCA, in the future, may help to overcome this challenge.

As discussed earlier, so far, TRoCA programme has been funded by the local funding bodies including local NGOs, National Public Health Management Centre (NPMC), Tabriz University of Medical Sciences and Ministry of

Health and Medical Education. However, TRoCA usually suffering of low salaries of its part-time personnel and lack of specific funds for prevention programmes. We therefore, at TRoCA, mainly rely on the funds received irregularly from NGOs. It has, therefore, been figured out in TRoCA that more financial support would be needed in the future activities for data collection and data analysis, and for monitoring purposes. This includes full-time individuals for data collection, analysis, and administrative works. More funds need to be, also, allocated to implement further interventional programmes for preventive purposes including programmes for high-risk groups, genetic counselling for those families who have a history of an anomaly in the family, etc.

7.8 Some Findings from TRoCA Data

A brief summary from the TRoCA data are presented below. More details of the epidemiological and clinical findings of the programme can be found in the peer-reviewed papers and reports published elsewhere [4–19]:

- A total of 305,504 births were registered in TRoCA since the registry established of which 302,156 (98.9%) were live births and 3348 (1.1%) stillbirths. During this period, 7830 cases with a primary diagnosis of congenital anomaly were ascertained, representing an overall prevalence rate of 271.2 per 10,000 births (95% CI: 265.4–276.9).
- Most of the infants born with birth defects were male (61.3%).
- The majority of infants were born live (89.6%).
- First cousin marriage was reported in 12.5% of the families with an affected child.
- Main type of delivery in infants born with congenital anomalies was caesarean section delivery (49.4%).
- Anomalies of genitourinary tract (45.7 per 10,000 births, 95% CI: 43.3–48.2), limb defects (40.07 per 10,000 births, 95% CI: 37.8.5–42.4), nervous system anomalies (36.7 per 10,000 births, 95% CI: 34.5–38.9), digestive defects (25.9 per 10,000 births, 95% CI: 24.1–27.8), and congenital heart diseases (34.3 per 10,000 births, 95% CI: 32.1–36.4) accounted for 77% of birth defects in the TRoCA region of the registry. The total prevalence of groups of congenital anomalies for the period of 2000–22 is shown in Fig. 7.1.
- There is a steady noticeable increase in the occurrence of congenital anomalies in the region over time ranging from 104.6 per 10,000 births (95% CI: 90.03–119.3) in 2000 to 628.3 per 10,000 births (95% CI: 556.2–700.5) in 2022 (Fig. 7.2). The specific prevalence rate of most categories of congenital anomalies showed no consistent downward or upward trend over the time from 2000 to 2022.
- Over the years of activities of the TRoCA registry, early records (2000–10) show that nervous system and genitourinary tract anomalies were the most frequent defects while later data (2011–21) indicated that genitourinary tract anomalies, limb defects, and heart anomalies are the most common defects. The frequency figures of the limb defects are almost the same in two periods of 2000–10 and 2011–21 in the TRoCA region (Fig. 7.3).
- The median survival time for those with congenital anomalies, who died, was 3 days (95% CI: 2.6–3.4). Mortality rate was 4.7 per 1000 live and still births (95% CI: 4.5–4.9), and there were 12.5 foetal deaths per 100 live births of congenital anomalies. Case fatality rate was estimated 18.5% (95% CI: 17.6–19.4) of birth defects.
- The proportions of all live-born infants with congenital anomalies surviving up to the first, second, and fourth weeks were 62% (95% CI: 56.1–67.9), 54% (95% CI: 46.2–61.8), and 45% (95% CI: 33.2–56.8), respectively.
- More than 3% (95% CI: 2.7–3.6) of congenital anomalies was medically terminated during hospital stay.
- The age of marriage, and consequently, the maternal and paternal age has increased over the past years in the region as an influencing

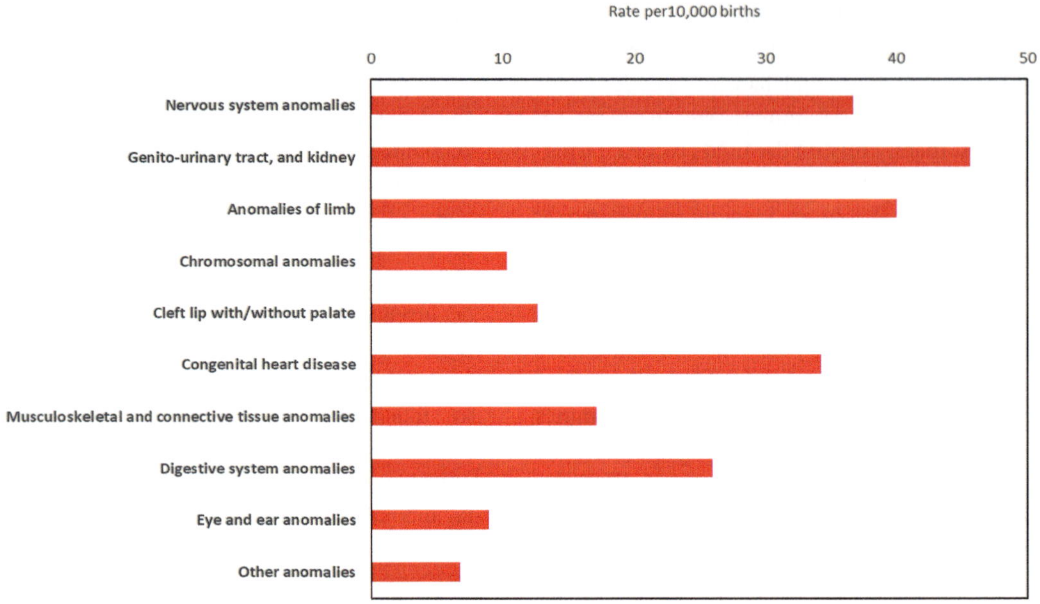

Fig. 7.1 Total prevalence of congenital anomalies (TRoCA, Iran)

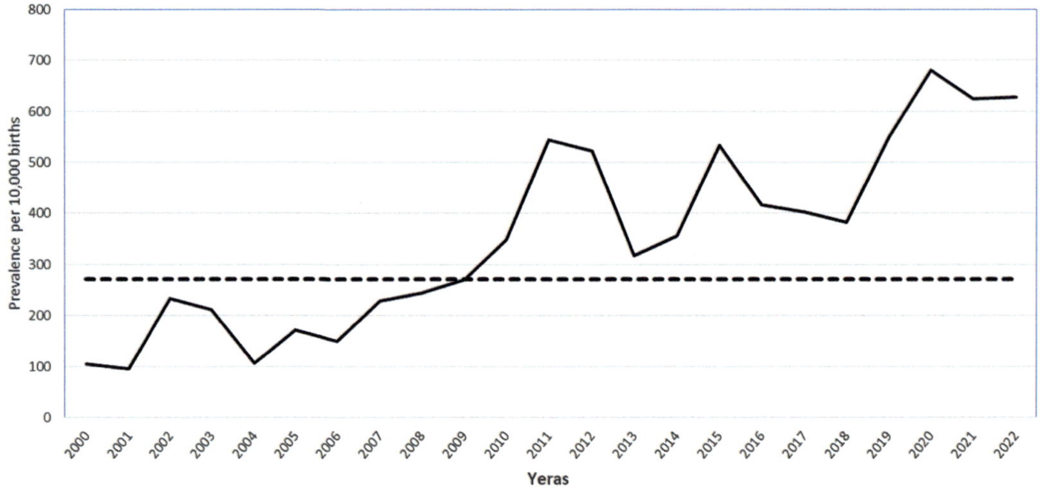

Fig. 7.2 Time trend of prevalence of congenital anomalies (TRoCA, Iran)

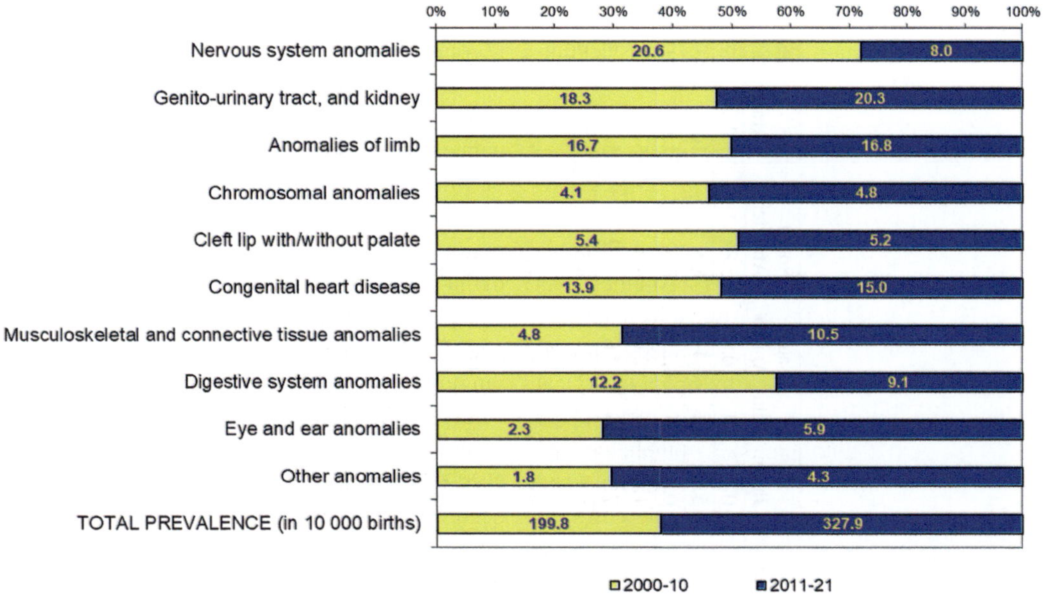

Fig. 7.3 Comparison of the defects proportion (%) between 2000–2010 and 2011–2021

factor on the occurrence of congenital and genetic disorders.

- Maternal ages fewer than 18 and over 35 years increased the risk of 2-month mortality of children born with congenital anomalies by 60% with an estimated risk of 1.60 (Odds Ratio = 1.60, 95% CI: 1.03–2.46).
- The mortality odds ratio for children born to mothers under 18 was estimated 1.65 (95% CI: 1.52–1.78), and the same risk for mother aged 35 or more was 1.70 (95% CI: 1.09–2.64).
- Diagnostic techniques were fully available in the study area both in public and private sectors. Screening of congenital hypothyroidism and thalassemia has been successfully performed across the country by the Ministry of Health over the past years.
- Some initial screening programmes have been proposed by the Ministry of Health to implement in the near future for neural tube defects, Down syndrome, and phenylketonuria.
- The high cost of genetic services does not allow many people to get access to the community genetic services despite their needs.
- Forty-eight percent of termination of pregnancies for birth defects was performed before the 18th week of pregnancy in the region.
- Consumption of folic acid prevented neural tube defects by 79% (Odds Ratio = 0.21, CI 95%: 0.12–0.40).
- Data indicated that 59.1% of children with congenital heart diseases were not identified at birth in the region covered by TRoCA programme.
- The accuracy rate of family physicians in case detection and diagnosis of congenital anomalies in rural areas in the TRoCA region is more than 98%.
- Table 7.1 shows the prevalence of selected congenital anomalies in 13 ICBDSR registries in the world. The rate of total limb reduction defects in TRoCA is almost nine times, in average, higher than that of other regions.
- International investigations compared the occurrence rates, mortality and survival pattern of selected groups of anomalies among various registries including TRoCA in the world. The details were presented in part one of this book [20–28].

Table 7.1 Prevalence (in 10,000 births) of selected congenital anomalies in 13 ICBDSR registries[a]

Selected congenital anomalies	Australia:WARDA[b]	Canada	France: Paris	Germany: Saxony-Anhalt	India	**Iran: Tabriz**	Italy: Tuscany	Japan	Malta	Saudi Arabia	South America[c]	Ukraine	USA: Texas
Anencephaly	5.5	1.2	4.9	2.0	11.3	**10.7**	1.5	0.7	2.9	10.5	5.8	7.0	2.7
Spina bifida	5.7	3.2	5.2	5.2	10.3	**1.4**	3.9	5.6	7.2	8.4	8.5	10.4	3.9
Encephalocele	2.1	0.7	2.2	1.4	2.1	**1.0**	1.0	0.5	1.5	4.2	3.3	1.8	1.0
Microcephaly	3.8	3.7	2.4	11.7	1.1	**4.5**	0.6	1.6	6.4	12.6	4.8	5.9	12.9
Hydrocephaly	7.7	4.7	14.4	4.6	8.4	**13.8**	3.4	7.8	1.5	12.6	17.2	5.3	7.2
Cleft palate without cleft lip	10.2	6.4	6.4	5.9	1.4	**8.4**	3.3	5.1	12.6	4.2	4.5	7.2	5.9
Cleft lip with or without cleft palate	10.8	9.4	8.2	12.1	4.0	**11.9**	5.2	21.7	10.7	13.6	11.7	7.8	10.4
Undescended testis	39.5	34.4	NR[d]	4.7	0.9	**9.8**	6.6	NR	NR	NR	9.8	31.6	13.7
Hypospadias	34.6	27.5	17.3	6.9	1.7	**19.8**	13.9	5.2	28.1	47.2	9.2	2.8	16.6
Limb reduction defects	7.0	3.3	6.1	7.7	2.1	**51.0**	4.2	3.8	6.3	6.3	7.6	4.8	6.0
Polydactyly	12.1	13.2	1.9	5.5	2.5	**8.0**	1.1	6.7	16.5	9.4	3.1	4.1	4.0

[a]Data based on the latest public report (2014) published by the International Clearinghouse for Birth Defects, Surveillance and Research (ICBDSR). Registries in this table are listed in alphabetic order. For the details of the populations covered in each registry, please refere to the ICBDSR website at: <www.icbdsr.org>

[b]Western Australian Register of Developmental Anomalies

[c]Latin American Collaborative Study of Congenital Malformations

[d]Not Reported

7.9 Further Achievements and Outputs

In addition to the achievements described above in relation to the TRoCA objectives, the programme has also been successful in the following areas:

One of the major aims of TRoCA (as a pilot programme) was to provide baseline information and framework to set up a nationwide registry of birth defects in the country. As a result, the first national registry of congenital anomalies was then established in 2012 in Iran. The programme is called Epidemiology and Surveillance of Congenital Anomalies in Iran (ESCAI). The main aims of this programme are to register the occurrence of selected birth defects in the whole country and to make valid data available to policy makers.

Tabriz Foundation for Public Health Genetics (TFPHG) was established in the area later in 2012 as an NGO. The principal aims of the foundation are to register some groups of birth defects and to implement control and preventive measures for genetic disorders and congenital anomalies in the region including genetic services to the families who have a history of an anomaly in the family, and preconceptional programmes for young couples.

Both the TFPHG and ESCAI programmes will be described in detail later in this section.

Six books (all in Persian) plus 10 peer-reviewed papers are published in relation to the aims of TRoCA.

TRoCA programme was awarded (2020) the Down syndrome research prize and medal for the achievements in the field of birth defects in the Eastern Mediterranean region by the World Health Organization.

7.10 Conclusions

A surveillance system for congenital anomalies (TRoCA) was set up in Tabriz University of Medical Sciences, Iran, in 2000 and running now for 23 years. The aims of TRoCA included documenting some epidemiological features of congenital anomalies in the population, providing information for healthcare needs for health service planners, clinicians, genetic counselling, and families, generation of data for research purposes, and implementing control and prevention measures of congenital anomalies in the region. Anecdotal and evidence-based reports indicate that the TRoCA programmes have been successful in most of their objectives over the past 23 years since established.

Although there is relatively a methodological consistency over time in the TRoCA programme, improvements in antenatal screening procedures and prenatal diagnosis technologies, changes in case ascertainment methods and sources of information, misclassification, errors in abstraction, coding and data entry should be considered as possible underlying contributing factors in seeking the time trends and increased prevalence of anomalies in the region. Maternal age, the rising proportion of older mothers and some other changes in the demographic pattern in the population, and environmental factors (including the role of preconceptional vitamin supplementation and folic acid) might too have contributed substantially to the time trend and prevalence of defects in the TRoCA region. It is also likely that existence of some specific local risk or causal factors for high occurrence of anomalies in the region cannot also be ruled out that needs further investigations.

The TRoCA programme is now a member of the International Clearinghouse for Birth Defects Surveillance and Research, a world affiliate member of the European network of registries for the epidemiologic surveillance of congenital anomalies, member at the International Federation for Spina Bifida and Hydrocephalus, and a member of the Society for Birth Defects Research and Prevention. TRoCA programme was awarded (2020) the Down syndrome research prize and medal for the achievements in the field of birth defects in the Eastern Mediterranean region by the World Health Organization.

The main challenges faced by TRoCA related to the reliance of the programme on part time staff for registry works, short-term and provisional funds, and difficulties in applying for international grants because of the political and financial sanction on many of the Iranian institutions over the past years.

References

1. Statistics & Administration Organization. 2023. https://www.amar.org.ir/Portals/0/Files/baravord/last26021401.xlsx?ver=MPny2GnWKDiZyT_-DE56DQ%3d%3d. Accessed 15 July 2023.
2. World Health Organization. International classification of diseases (ICD). 2023. https://icd.who.int/en.
3. Lee ML, Teutsch SM, Thacker SB, St. Louis ME. Principles and practice of public health surveillance. Oxford: Oxford University Press; 2010.
4. Dastgiri S, Heidarzadeh M, Dastgiri A. Tabriz registry of congenital anomalies: a report of 10 years of monitoring birth defects in Iran. Congenit Anom. 2013;53(2):98–9.
5. Stone D, Dastgiri S, Heidarzadeh H, Abdollahi HM, Imani S, Maher MHK. Uses, limitations, and validity of a registry of congenital anomalies in Iran: a critical review. J Environ Public Health. 2017;2017:6972617.
6. Tarighat F, Golshan E, Dastgiri S. Prevalence of congenital anomalies in the northwest of Iran. Depiction Health. 2021;12(4):417–25.
7. Dastgiri S, Sheikhzadeh Y, Dastgiri A. Monitoring of congenital anomalies in developing countries: a pilot model in Iran. Stud Health Technol Inform. 2011;164:157–61.
8. Dastgiri S, Kalankesh LR, Heidarzadeh H, Tajahmad A, Rezaian E. A new registry of congenital anomalies in Iran. J Registry Manag. 2010;37(1):27–9.
9. Abdolahi HM, Maher MHK, Afsharnia F, Dastgiri S. Prevalence of congenital anomalies: a community-based study in the northwest of Iran. ISRN Pediatr. 2014;2014:920940.
10. Dastgiri S, Bonyadi MJ, Mizani T. Epidemiology of neuro-genetic disorders in northwestern Iran. Neurosciences. 2012;17(2):171–2.
11. Dastgiri S, Salehpoor F, Bigdeli A. An unusual time pattern of central nervous system anomalies in Northwestern Iran. Neurosciences. 2010;15(4):294.
12. Dastgiri S, Imani S, Kalankesh L, Barzeghar M, Heidarzadeh M. Congenital anomalies in Iran:a cross-sectional study on 1574 cases in the north-west of country. Child Care Health Dev. 2007;33(3):257–61.
13. Dastgiri S, Taghizadeh M, Heidarzadeh M. Early diagnosis and screening of congenital cardiac anomalies. Cardiol Young. 2011;21:194–6.
14. Mashhadi Abdolahi H, Kargar Maher MH, Karamouz M, Afsharnia F, Khosroshahi H, Dastgiri S. How accurate is diagnosis of congenital anomalies made by family physicians? Health Promot Perspect. 2014;4(2):158–64.
15. Mashhadi Abdolahi H, Kargar Maher MH, Karamouz M, Khosroshahi H, Dastgiri S. Reliability and validity of the march of dimes preconception/prenatal family health history questionnaire: the Persian version. Congenit Anom. 2016;56(3):107–11.
16. Bateni A, Dastgiri S, Mosaferi M, Heidarzadeh M. A review of spatial analysis procedures: an application for congenital anomalies registries. J Clin Res Gov. 2013;2:70–2.
17. Samadirad B, Khamnian Z, Hosseini MB, Dastgiri S. Congenital anomalies and termination of pregnancy in Iran. J Pregnancy. 2012;2012:1–4.
18. Hosseini MB, Khamnian Z, Dastgiri S, Samadi Raad B, Ravanshad Y. Folic acid and birth defects: a case study. J Pregnancy. 2011;2011:370458.
19. Heidari F, Dastgiri S. Maternal age and birth defects in Iran. Eur J Gen Med. 2014;11(2):130–1.
20. Kancherla V, Sundar M, Tandaki L, Lux A, Bakker MK, Bergman JE, et al. Prevalence and mortality among children with anorectal malformation: a multi-country analysis. Birth Defects Res. 2023;115(3):390–404.
21. Gili JA, López-Camelo JS, Nembhard WN, Bakker M, de Walle HEK, Stallings EB, et al. Analysis of early neonatal case fatality rate among newborns with congenital hydrocephalus, a 2000–2014 multi-country registry-based study. Birth Defects Res. 2022;114(12):631–44.
22. Bell JC, Baynam G, Bergman JEH, Bermejo-Sánchez E, Botto LD, Canfield MA, et al. Survival of infants born with esophageal atresia among 24 international birth defects surveillance programs. Birth Defects Res. 2021;113(12):945–57.
23. Politis MD, Bermejo-Sánchez E, Canfield MA, Contiero P, Cragan JD, Dastgiri S, et al. Prevalence and mortality in children with congenital diaphragmatic hernia: a multicountry study. Ann Epidemiol. 2021;56:61–69.e3.
24. Nembhard WN, Bergman JEH, Politis MD, Arteaga-Vázquez J, Bermejo-Sánchez E, Canfield MA, et al. A multi-country study of prevalence and early childhood mortality among children with omphalocele. Birth Defects Res. 2020;112(20):1787–801.
25. Bakker MK, Kancherla V, Canfield MA, Bermejo-Sanchez E, Cragan JD, Dastgiri S, et al. Analysis of mortality among neonates and children with Spina bifida: an international registry-based study, 2001–2012. Paediatr Perinat Epidemiol. 2019;33(6):436–48.
26. Goel N, Morris JK, Tucker D, de Walle HEK, Bakker MK, Kancherla V, et al. Trisomy 13 and 18-prevalence and mortality-a multi-registry population based analysis. Am J Med Genet A. 2019;179(12):2382–92.
27. Yu X, Nassar N, Mastroiacovo P, Canfield M, Groisman B, Bermejo-Sánchez E, Ritvanen A, et al. Hypospadias prevalence and trends in international birth defect surveillance systems, 1980–2010. Eur Urol. 2019;76(4):482–90.
28. Mc Goldrick N, Revie G, Groisman B, Hurtado-Villa P, Sipek A, Khoshnood B, et al. A multi-program analysis of cleft lip with cleft palate prevalence and mortality using data from 22 international clearinghouse for birth defects surveillance and research programs, 1974–2014. Birth Defects Res. 2023;115:980–97.

8 Tabriz Foundation for Public Health Genetics

Content

There are always waivers built into the health services by the governments. Those waivers however never functioning well enough because of the ever-increasing "budget shortage" problems in the country. For this reason, ministerial authorities are usually unable to fully support the services including the new field of community genetics activities. Even if they support, the sustainability is always in doubt. These, in general, will have considerable consequences on all aspects of the population health status. Many people might, therefore, be affected. Of them, children and women, for obvious reasons, are the most vulnerable groups.

As discussed in the earlier chapter, Tabriz Registry of Congenital Anomalies (TRoCA) has operated under the authority of Tabriz University of Medical Sciences. Until recently, the maintenance of the programme has become possible by the financial support from the following sources in the past: National Public Health Management Centre (NPMC), Ministry of Health and Tabriz University of Medical Sciences. Following the periods of short-term funding from governmental sources, the committee of the registry realized that TRoCA needs more funds, without relying to governmental sources, to be able to extend its activities, and to implement control measures for the prevention of congenital anomalies and genetic disorders in the community. Based on this innovative but challenging idea, a new Non-Governmental Organization (NGO) was established in the area later in 2012 called Tabriz Foundation for Public Health Genetics (TFPHG) as an organization that is neither a part of a government nor a conventional for-profit business.

The principal aims of the TFPHG are to assist TRoCA to register the birth defects and to implement control and preventive measures for genetic disorders and congenital anomalies in the region including genetic services to families who have a history of an anomaly in the family, and preconceptional programmes for young couples. The foundation is now working in a range of activities in the region including neonatal screening for some congenital and genetic disorders (i.e. inherited children deafness), education, genetic consultation and diagnostic genetic tests for prevention, and consultation for the therapeutic termination of pregnancies for birth defects.

TFPHG was officially registered as a self-financed NGO in the region with no need to the governmental resources. It has been recognized and acknowledged by the Ministry of Health.

S. Dastgiri, *Congenital Anomalies in Iran*, https://doi.org/10.1007/978-981-97-6274-3_8

According to the administrative rules, TFPHG acting in accordance with/under the public health policies regulated by the Ministry of Health and the State Welfare Organization. The foundation is a charity-based programme financed mainly by its own income, by non-governmental entities and individuals and, partially, by the State Welfare Organization. The details of the funding resources at TFPHG include:

- The committee of the NGO has discussed the importance of its prevention objectives to some organizations and rich individuals in the region convincing them to donate and fund the TFPHG for its purposes and activities. Fortunately, it has well been received by those organizations and individuals. They now fund TFPHG for some parts of its activities. This is a typical fund-raising procedure well-known in Iran.
- By providing preventive genetic services for individuals and families, TFPHG charges them at a minimum cost for these services.
- The activities performed at TFPHG have well been recognized and acknowledged by the State Welfare Organization in Iran. Therefore, the programme too receives donations from this governmental organization for some parts of its activities.

The steering committee members at TFPHG include 20 professionals from various disciplines involved in the community health in the region. TFPHG has also recruited two lawyers for managing the legal matters raised as bio-ethical and societal issues in implementing the community genetic services.

8.1 Conclusion

It is believed that TFPHG, as a Non-Governmental Organization without relying to the governmental resources, creates an innovative and realistic frame work for screening and diagnostic procedures aiming mostly at primary and tertiary prevention levels of genetic disorders and congenital anomalies. In the future, the framework and action plans of TFPHG may provide a pilot model and insights into the epidemiology and potential for prevention and control of congenital anomalies in the whole region and similar areas.

9 Epidemiology and Surveillance of Congenital Anomalies in Iran

Contents

As reported in the first chapter of this section, there has been no national evidence-based data of the occurrence of congenital anomalies in Iran until the TRoCA was set up in the northwest of the country. In this chapter, the nationwide situation of congenital anomalies in Iran will briefly be reported from the national registry programme of birth defects. The national programme collects data on the occurrence of congenital anomalies in Iran producing prevalence rates of defects and making geographical comparisons across the country. The ultimate aims of the data gathered in this registry would potentially enable aetiological studies and service provision investigations to response to the healthcare needs of high-risk population.

Iran has an area of 1,648,195 km^2 with an estimate of 84.1 million population, as estimated for 2021, located in the Middle East. Iran is a big and very diverse multicultural country in the region consisting of many ethnic minorities including Persians, Azeris, Kurds, Lurs, Balochs, Arabs, Turkmens, Gilaks, Talyshs, Mazanis, and Qashqais. In terms of religious groups, Muslims (Shia, Sunni, and other Islamic subgroups), Jews, Christians, Zoroastrian, and some other religious minorities live in various places in the country. Consanguineous marriages are legally permitted in the country and are common in most ethnic and religious groups and communities. The country is divided into 31 provinces. Iran hosts one of the largest refugee populations in the world mostly from Afghanistan and Iraq [1]. The country borders are with Armenia, Azerbaijan, Nakhchivan, Caspian Sea, Turkmenistan, Afghanistan, Pakistan, the waters of the Persian Gulf and the Gulf of Oman, Iraq, and Turkey.

There has been no evidence-based data available on the occurrence of birth defects in Iran until 2000 when Tabriz Registry of Congenital Anomalies (TRoCA) was launched to investigate the epidemiology of congenital anomalies in the northwest region of the country [2]. One of the main aims of TRoCA, as discussed before, was to provide baseline information and framework to set up a nationwide registry of birth defects in the country. As a result, the first national registry of congenital anomalies was then established in 2012 in Iran. The programme is called Epidemiology and Surveillance of Congenital Anomalies in Iran (ESCAI). The principal aims of ESCAI are to register the occurrence of

S. Dastgiri, *Congenital Anomalies in Iran*, https://doi.org/10.1007/978-981-97-6274-3_9

selected birth defects in the whole country and to make valid data available to policy makers, research purposes, clinicians, and genetic counsellors.

Data collection, process, analysis, and evaluation of ESCAI are based on the guidelines suggested by the International Clearinghouse for Birth Defects Surveillance and Research (ICBDSR), and European Network of Registries for Congenital Anomalies [3, 4]. The guidelines have been validated for regional settings by TRoCA programme. ESCAI is located in the capital city, Tehran, and funded by the Iran Ministry of Health.

All children in the maternity hospitals across the country are examined by a gynaecologist, obstetrician, or midwives at birth. In some areas where the facilities and personnel allow, neonatologists, paediatricians, and other relevant clinical professionals are present at the time of delivery, as needed. Newborns are normally followed up until hospital discharge for general health status, maturity, and some obvious clinical conditions including birth defects. Congenital anomalies are defined based on the coding system of the International Classification of Diseases (ICD) according to the primary diagnosis of anomaly based on the list of defects provided by TRoCA programme.

Total prevalence is calculated by dividing the numerator (registered cases of congenital anomalies) by the relevant denominator (total births). An infant/foetus with more than one anomaly is counted once only in the numerator.

The ESCAI programme has registered all births (an average of 1,000,000 for every year) in the country since 2013. The prevalence rate of major selected groups of anomalies in the country including congenital heart diseases, digestive system anomalies, genitourinary tract, kidney, musculoskeletal and connective tissue anomalies are annually reported for every province and city across the country. There are always noticeable increasing or decreasing time pattern, and unusual occurrence for some specific groups of defects in some regions.

In the ESCAI, the epidemiological profile of birth defects is considerably underestimated compared to the global figures and to the data reported from a regional registry of anomalies TRoCA. Accordingly, based on unpublished reports, the musculoskeletal and connective tissue anomalies four times, genitourinary tract, and kidney defects five times, digestive system anomalies nine times, and congenital heart diseases 15 times are seemingly less estimated by ESCAI as compared to the TRoCA registry and international data [2–4]. It has been figured out that in this newly established nationwide programme, the condition of antenatal screening procedures, detection and ascertainment methods of defects, lack of professional medical staff in many of small cities and regions, variable validity of the data and low power of the programme to find rare defects are now of main limitations and weaknesses in the surveillance of the congenital anomalies across the country. Hence, target-oriented training of professional and technical staff involved in maternity and neonatal facilities may help ESCAI to produce true estimates of birth defects for the country in the future.

As discussed before, the evaluation and assessing the major components of a standard surveillance system for validating, monitoring the strengths, shortcomings and weaknesses of the programme, and to assure the quality (accuracy and completeness) of the data must be of principal priorities in every registry including ESCAI. Therefore, the following evaluation criteria will have to be implemented in the programme in the years ahead: simplicity, flexibility, acceptability, reliability, utility, sustainability, timeliness, sensitivity, and representativeness [5, 6]. As a result, the future reports from ESCAI registry will be able to document much more accurate estimates of the occurrence of birth defects, longer time trend analysis, and data on more groups of congenital anomalies. Until then, TRoCA will still remain as the main representative source of the occurrence pattern of congenital anomalies in the country and similar areas.

9.1 Conclusions

The data gathered by ESCAI potentially enables service provision to response to the healthcare needs. This national registry has apparently underestimated the occurrence of the congenital anomalies in the country compared to the global figures and to the data reported from the regional registry of anomalies in the country, TRoCA. Target-oriented training of professional staff involved in maternity and neonatal facilities may help to produce true estimates of birth defects in the years ahead.

References

1. Statistics & Administration Organization. 2023. https://www.amar.org.ir/Portals/0/Files/baravord/last26021401.xlsx?ver=MPny2GnWKDiZyT_-DE56DQ%3d%3d.
2. Stone DH, Dastgiri S, Heidarzadeh M, et al. Uses, limitations, and validity of a registry of congenital anomalies in Iran: a critical review. J Environ Public Health. 2017;2017:6972617.
3. International Clearinghouse for Birth Defects Surveillance Research (ICBDSR). Birth defects prevalence. 2023. http://www.icbdsr.org/resources/annual-report.
4. European network of registries for the epidemiologic surveillance of congenital anomalies (EUROCAT). 2023. https://eu-rd-platform.jrc.ec.europa.eu/eurocat/eurocat-data/prevalence_en.
5. Dastgiri S, Heidarzadeh M, Dastgiri A. Tabriz registry of congenital anomalies: a report of 10 years of monitoring birth defects in Iran. Congenit Anom (Kyoto). 2013;53:98–9.
6. Lee ML, Teutsch SM, Thacker SB, St. Louis ME. Principles and practice of public health surveillance. Oxford: Oxford University Press; 2010.

Part III

Disability at a Glance

10

Contents

The definition of disability and impairments, the epidemiological occurrence in the world will briefly be reviewed in this section. The status of the measurement in the population and epidemiological profile of the disability in Iran and in the TRoCA region are discussed and presented separately later in the section.

10.1 What Is Disability?

In general, there is no a systematic definition of disability among countries. The occurrence rates might therefore vary remarkably between different parts of the world. The World Health Organization provided, in 2001, a framework definition based on the details of the anatomical structure, physiological function and functional tasks, daily life activities, and social participation of the individuals experiencing disability [1].

Disability is broadly defined as any condition of the body or mind (including vision, movement, thinking, remembering, learning, communicating, hearing, mental health, and social relationships) that causes activity limitation (i.e. seeing, hearing, walking, or problem-solving), participation restrictions in the normal daily life (i.e. working and engaging in social activities), difficulties in interaction with others around, and impairments (including both structural and functional). Impairment is defined as the absence or variation in the component of the body structure or functions (i.e. any brain damage that may cause multiple sclerosis, difficulty in joints movements, and difficulty of vision or hearing), or mental functioning.

In terms of aetiological aspects, the disabilities and impairments can originally be related to congenital abnormalities that are present at birth and later in life (i.e. single genes disorders, disorders of chromosomes, mother's harmful exposure during pregnancy), developmental conditions (i.e. autism spectrum disorder and attention-deficit/hyperactivity disorder), injuries (i.e. traumatic brain and spinal cord injuries), and chronic or progressive conditions including any disease and illness that may cause various problems in the body or mental functions (i.e. diabetes, muscular dystrophy, and multiple sclerosis) [2].

S. Dastgiri, *Congenital Anomalies in Iran*, https://doi.org/10.1007/978-981-97-6274-3_10

10.2 Occurrence in the World

An overview and key facts of the current status of the disability prevalence in the world and related factors and conditions are briefly outlined by the World Health Organization as shown in the box below [3]:

- *An estimated 1.3 billion people experience significant disability. This represents 16% of the world's population, or 1 in 6 of us.*
- *Some persons with disabilities die up to 20 years earlier than those without disabilities.*
- *Persons with disabilities have twice the risk of developing conditions such as depression, asthma, diabetes, stroke, obesity, or poor oral health.*
- *Persons with disabilities face many health inequities.*
- *Persons with disabilities find inaccessible and unaffordable transportation 15 times more difficult than for those without disabilities.*
- *Health inequities arise from unfair conditions faced by persons with disabilities, including stigma, discrimination, poverty, exclusion from education and employment, and barriers faced in the health system itself.*

As presented in the table, conditions related to the disability shows a remarkable impact on the community health across the globe notable 1 in 6 of the world population suffering from various disabilities and impairments with significant impact on their life course needing a huge amount of care attention and medical intervention by families and public health authorities. As for time trends, the prevalence of disability is growing over decades for various reasons and influencing factors including the ageing pattern of the population, increase in chronic diseases and health conditions, traffic accidents, natural disasters, conflicts, diet, and substance abuse. The earlier report, 1970, of the World Health Organization estimated 10% of the global population as the prevalence of disabilities where the same figure increased to 15% in 2011 and later years (Fig. 10.1) [4].

The Global Burden of Diseases (GBD), Injuries, and Risk Factors Study is the most recent assessment of the incidence, prevalence, and years lived with disability (YLDs) for 354 non-fatal causes and impairments in 195 countries and territories for the years 1990, 2007, and 2017 [5]. According to the findings of this comprehensive systematic analysis of the global burden of diseases and conditions, the time percentage change in the age-standardized rates was estimated −2.4% (1990–2007) and −1.5% (2007–2017), +7.1% (1990–2007) and +2.1% (2007–2017), and 0.0% (1990–2007) and −0.5% (2007–2017) for impairments of vision loss, developmental intellectual disability and hearing loss, respectively (Table 10.1). However, the details of the findings of this systematic analysis indicated that different age groups, sexes, and geographical areas across the globe experiencing various burdens over nearly three decades of investigation for those conditions and impairments [5].

Over the few decades due to various reasons, disability-related disorders have increasingly contributed to the global burden of diseases and health expenditure in the world population. Greater attention and more actions are therefore required by policy and decision-makers in terms of investment in public and private sectors for effective intervention strategies, research programmes, and improvement of the quality life of disabled individuals [6].

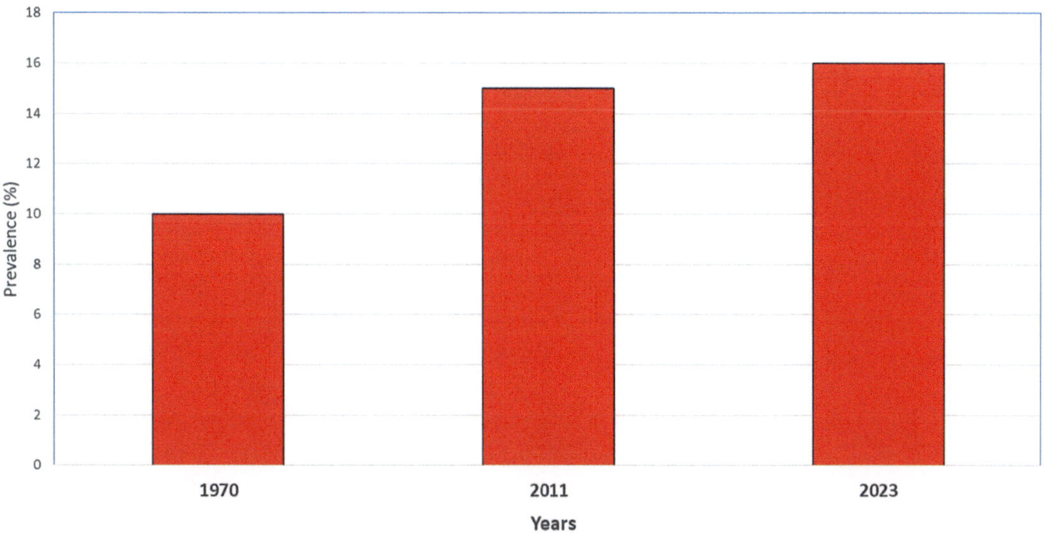

Fig. 10.1 Prevalence (%) of disabilities in the world population

Table 10.1 Global prevalence, and YLDs for 2017; percentage change of YLD counts; and percentage change of age-standardised YLD rates for 1990–2007 and 2007–2017 for three impairments [3]

	Prevalence (thousands)	YLDs (thousands)				
Impairments	2017 counts	2017 counts	Percentage change in counts, 1990–2007	Percentage change in counts, 2007–2017	Percentage change in age-standardised rates, 1990–2007	Percentage change in age-standardised rates, 2007–2017
Hearing loss	1,428,450.1	39,454.3	40.0	22.4	0.0	−0.5
Vision loss	1,339,970.3	34,125.4	38.3	22.2	−2.4	−1.5
Developmental intellectual disability	188,585.1	25,301.2	30.2	12.9	7.1	2.1

References

1. World Health Organization. International Classification of Functioning, Disability and Health (ICF). 2023. https://www.who.int/standards/classifications/international-classification-of-functioning-disability-and-health.
2. Center for Disease Control and Prevention. Disability and health promotion. 2023. https://www.cdc.gov/ncbddd/disabilityandhealth/index.html.
3. World Health Organization. Disability fact sheet. 2023. https://www.who.int/news-room/fact-sheets/detail/disability-and-health.
4. World Health Organization. World report of disability. Geneva; 2011.
5. GBD 2017 Disease and Injury Incidence and Prevalence Collaborators. Global, regional, and national incidence, prevalence, and years lived with disability for 354 diseases and injuries for 195 countries and territories, 1990–2017: a systematic analysis for the Global Burden of Disease Study 2017. Lancet. 2018;392(10159):1789–858.
6. GBD 2019 Diseases and Injuries Collaborators. Global burden of 369 diseases and injuries in 204 countries and territories, 1990–2019: a systematic analysis for the Global Burden of Disease Study 2019. Lancet. 2020;396(10258):1204–22.

State Welfare Organization of Iran 11

Contents

Before the Iran revolution (1979), the following major organizations and societies used to provide services to vulnerable individuals and families at various levels in the country:

- National Support Association for Children
- Supporting Society for Homeless Children
- Association to Support Disabled
- National Deaf Welfare Organization of Iran
- National Blind Welfare Organization of Iran
- Shahanshahi Organization for Social Services
- Rural Culture Houses
- Welfare and Education Organization for Children and Teenagers
- Educational and Training Units of the Municipalities
- National Family Welfare Organization of Iran

In order to fulfil the provisions and services of those entities in planning, coordination and supervision, the State Welfare Organization of Iran (SWO) was established in 1980 aiming mainly to support homeless families, abused children and deprived people, asylum-seeking girls and women (fugitives), orphans, physically disabled and mentally handicapped individuals, spinal cord injuries, elderly people, and other similar vulnerable individuals and families.

Currently, the SWO is actively involved in providing care and services including: kindergartens, 24-h care centres for homeless and street children, health centres for girls and women, clinics and emergencies for community nursing and psychology, social counselling and social psychology service centres, rehabilitation centres for disabled individuals, vocational training centres for the disabled, rehabilitation and care centres for the elderly, rehabilitation and treatment centres for chronic mental patients, genetic counselling services, counselling centres for women seeking divorce, empowerment centres for self-employed and female-headed households, suicide prevention centre, prevention programmes for disabilities, prevention programmes for social

S. Dastgiri, *Congenital Anomalies in Iran*, https://doi.org/10.1007/978-981-97-6274-3_11

harms, treatment and rehabilitation centres for drug addicts (i.e. harm reduction and prevention of the spread of AIDS). The SWO is also closely collaborating with non-governmental and charity associations and institutions that work in line with the goals of SWO across the country for maintaining public participation and voluntary services. The details of the specific missions of the State Welfare Organization can be found on its website [1].

The SWO comprises three main departments including Prevention, Socio-Cultural Affairs, and Rehabilitation.

11.1 Prevention Department

Designing, monitoring, and implementing activities to prevent disabilities and harm reduction at individual, family, and community levels are of activities of this department. The following goals are specifically pursued by the department of prevention:

- Identifying the causes of disabilities and social harms
- Planning for the reduction of the incidence of disabilities
- Modifying the environmental harm factors (i.e. biological, psychological, and sociological)
- Creating and strengthening of logistic resources for prevention purposes

The activities for those objectives are implemented by four sub-divisions at this department: Prevention of Disabilities (i.e. screening for deafness and visual impairments, genetic disorders), Social Harms Reduction (i.e. skill-based training workshops for interpersonal relationships, cognitive and physical matters, effective communication skills, creative thinking skills, critical thinking, problem-solving, decision-making, and coping skills to deal with stress and negative emotions at various levels from children to adults audience), Prevention and Treatment of Addiction (i.e. addiction prevention in schools, families, workplace, detoxification, and empowerment of addicts), and Psychological Counselling Services (i.e. 24-h face-to-face and telephone psychological counselling for emotional, family, and social conflicts).

11.2 Department of Socio-Cultural Affairs

The activities of this department are mainly focused in three major disciplines for:

- Children and Adolescents (i.e. supporting and taking care of orphans, children from poor and marginal areas, unsupervised children providing various services to the individuals from under primary school age to adolescents)
- Family and Women Empowerment (i.e. supporting female-headed families for literacy matters, vocational training and job preparation, life skills, free insurance, housing)
- Social Victims (i.e. taking care of street children, abused persons, supporting individuals exposed to domestic violence in safe houses, programmes to reduce divorce and suicide plus a 24-h social emergency helpline)

11.3 Rehabilitation Department

The main aim of the rehabilitation department of SWO is to plan for integrating the disabled individuals as normal and active citizens in the community through fulfilling their normal life plus equal opportunities inside the family so that a disabled person can enjoy the same rights as other members of the society do. To achieve this end, the following activities are pursued for disabled people in the rehabilitation department of SWO:

- Adaptation of the urban environment and public places
- Adaptation of the residential environment
- Psychology counselling services
- Preschool training programmes for children with severe disabilities
- Providing services and financial aids

- Providing health care and nursing services for special groups (i.e. physically disabled, elderly, deaf, blind, and mentally retarded individuals)
- Day and night housing services for special groups (i.e. physically disabled, elderly, deaf, blind, and mentally retarded individuals)
- Advanced psychological and rehabilitation services (i.e. physiotherapy, speech therapy, occupational therapy, audiology, orthopaedics surgery, psychology)
- Training for social adaptation and communication skills
- Supporting disabled individuals to take part in the world sport events including paralympic games and other world competitions
- Referring to disability cases requiring more specialized services to higher medical and care services
- Setting up the local and regional rehabilitation councils at the village and district level

11.4 Iranian Red Crescent Society

In addition to the State Welfare Organization of Iran, the Iranian Red Crescent Society is also, by some means, involved in providing various types of medical and healthcare services to disabled individuals and families in the field of rehabilitation, surgical corrections and humanitarian aids. The society was originally established in the country, 1905, as a branch of the International Committee of the Red Cross mainly for humanitarian activities inside and outside the country [2].

References

1. State Welfare Organization of Iran (SWO). 2023. https://www.behzisti.ir.
2. Iranian Red Crescent Society. 2023. https://en.rcs.ir.

12 Disability in Iran

Contents

12.1 Background

The majority of the world's population lives in less-developed countries where, because of limited resources, there are challenges and a great paucity of epidemiological profile of disability and impairments at the community level, and lives and quality of life of disabled individuals are largely unknown to us. Poverty in many of those countries likely exacerbates the daunting challenges of disability in the population. The lack of information about disability should draw the attention of public health authorities as it is expected that one in six of individuals suffering of significant disability according to the world Health Organization [1]. Data collection, improvement of the diagnostic and monitoring systems, and quality investigations on disabilities in those developing areas is, therefore, of great importance. This will enable public health system to have access to the valid data and accurate information, effective management, planning and evaluation of the prevention programmes, policy making, and estimation the essential needs for future and forecast and scenarios. Cross-sectional and prevalence studies, in a regular periodic basis, are of initial requirements to achieve those holistic goals for the management of disabilities in the population.

There is no registry to estimate the true figures of various types of the disabilities and impairments in Iran. Apart from some sporadic studies on various aspects of disabilities, there has been no comprehensive accurate data on the occurrence of disabilities and impairments in the country. For the first time, a number of limited questions were included in the national census of the country in 2006. The census data included the frequency of some specific types of disability in the family by age, gender, and place of living (i.e. rural, city, district, province). In the census, there was no administrative and planning data, compilations, aetiology, organizations and entities involved, and the relevant cultural and social conditions for each type of disability and impairments.

S. Dastgiri, *Congenital Anomalies in Iran*, https://doi.org/10.1007/978-981-97-6274-3_12

12.2 Nationwide Occurrence of Disability: Census Data

The National Census of Population and Housing in Iran is conducted by Statistics Center of Iran on a regular basis. Exceptionally, the nationwide census programmes in the years 2006 and 2011 included some questions about disability at individual and household levels in the country. The total population in Iran was counted 70,495,782 and 75,149,669 in the years 2006 and 2011, respectively. The specific types of disability in the census data included visual, hearing, voice and speech, arm and hand, leg and foot, trunk, and intellectual conditions [2]. Later, using the same census data, some other studies were carried out by researchers. One of which was a secondary analysis on 2% of the census data that was carried out on a sample of 9848 disabled and 1,235,445 non-disabled individuals, plus 17,166 households with at least one disabled person and 298,741 households with no disabled individuals in the family [3, 4].

Table 12.1 compares the total and sex-specific prevalence rates of disability in two national census between 2006 and 2011 in Iran. The table shows both the rates based on the total number of disabilities counted (per 1000) in the population (17.84 for 2006, and 16.91 for 2011) and the number of individuals having at least one disability (14.36 for 2006, and 13.54 for 2011). In either conditions, the total rate of disability slightly decreased over the time between 2006 and 2011. It was found that, approximately, 14 (per 1000) of Iranian individuals had at least one type of disability in both census data. The household data showed that more than 5% of the families across the country had at least one disabled individual as the family member. The rates for males were considerably higher compared to the female rates in either censuses.

The prevalence for different types of disabilities in presented in Table 12.2. Leg and foot defects and intellectual disabilities were the most common type of disability, either in males and females, in the country in both period of time between 2006 and 2011. Similarly to the total prevalence of disability in the country, the figures for all types of disabilities were higher among men compared to women.

The national census data in 2006 and 2011 showed that about 20% of the disabled individuals suffering from more than single disability in the population. They might have two or three disabilities at the same time (Table 12.3).

Further secondary data analysis studied the impact of socioeconomic status, age, sex, education, employment, marital status, and residence condition on the prevalence of the different types of disability in Iran. More than 50% of the proportionally distribution of disabilities was observed in 15–44 age groups. The findings indicated that the prevalence of disability was higher in the rural area compared to the urban regions. Remarkable socioeconomic inequalities in the pattern of occurrence of disability in the country were found where the socioeconomic status of disabled people is considerably lower compared to the normal population. Female-headed households were at higher risk of having a disabled person in the family [3, 4].

Table 12.1 Prevalence of Disability (Iran National Census of Population and Housing Data)

	Prevalence (thousands)	Prevalence (thousands)
	Based on total number of disabilities	Based on individuals with at least one disability
Total (2006)	17.84	14.36
Male	22.20	18.04
Female	13.31	10.04
Total (2011)	16.91	13.54
Male	20.85	16.81
Female	12.91	10.21

Table 12.2 Prevalence of types of disabilities (Iran National Census of Population and Housing Data)

	Prevalence (thousands), 2006			Prevalence (thousands), 2011		
Disabilities	Total	Men	Female	Total	Male	Female
Visual	1.3	1.6	1.1	1.3	1.5	1.1
Hearing	1.3	1.5	1.1	1.3	1.5	1.2
Voice and speech	2.2	2.6	1.7	1.8	2.2	1.5
Arm and hand:						
Amputation	0.3	0.5	0.1	0.3	0.5	0.1
Defect	2.0	2.8	1.1	1.7	2.4	1.0
Leg and foot:						
Amputation	0.7	1.0	0.4	0.6	0.9	0.3
Defect	4.2	5.2	3.0	3.5	4.3	2.7
Trunk	1.7	2.0	1.4	1.8	2.2	1.5
Intellectual	4.2	5.1	3.3	4.4	5.3	3.5

Table 12.3 Frequency of disabilities in individuals (Iran National Census of Population and Housing Data)

	Percent, 2006			Percent, 2011		
Number of disabilities	Total	Men	Female	Total	Male	Female
One	81.0	81.9	79.5	80.8	81.5	79.6
Two	13.7	13.2	14.7	13.5	13.0	14.3
Three	5.3	4.9	5.8	5.7	5.5	6.1

12.3 Disability in the National Surveys

A number of national surveys were conducted over the past few decades in Iran aiming to document the population profile of a variety of health and diseases indicators mostly for mental health issues or some particular diseases in specific age groups (i.e. injuries, paediatric asthma). There are three major national health surveys carried out in 1990, 2000, and 2010 in which the occurrence pattern of major diseases and general health of the population in rural and urban households were investigated. These national health surveys were administered and supported by the governmental entities. The main aims of the national health surveys were to provide data on health at the national and provincial levels, especially on children and women situation in Iran, and to assist policy makers for designing effective strategies to promote health outcomes and equity in access to health care in the country. Although the surveys did not specifically gather the disability data from the population, an overall prevalence of 27.03 (per 1000) was estimated for the total number of the various types of disabilities in Iran [5–8].

There is a considerable difference between the rates reported by the national survey and census data. This might be due to the fact that census programmes cover the whole population regardless of age whereas national surveys investigated the occurrence of disability (and other conditions) in adult individuals usually older that the 18-years in the population. This indicates that the prevalence of various kinds of disability in adult population is expected to be higher compared to the whole population when all age groups are included.

12.4 State Welfare Organization Surveys on Disability and Impairments

As discussed in the earlier chapter, one of the main aims of the State Welfare Organization of Iran is to support and take care of the disabled and mentally handicapped individuals and vulnerable families. Accordingly, the Welfare Organization would identify the individuals with

Table 12.4 Frequency distribution of impairments in Iran (Iran State Welfare Organization Data) [9]

Impairments	Percent, 2018–23
Visual	19.60
Hearing	12.58
Voice and speech	7.99
Mental	13.46
Structure of brain and spinal cord	12.13
Movement	34.24

disabilities, working on the causes related to the occurrence of disabilities, implementing interventional and preventive programmes in individuals, family and community levels.

There have, on irregular basis, been some sporadic surveys on disabilities (and other conditions and diseases) performed by the Welfare Organization in various occasions in different parts of the country.

A comprehensive national cross-sectional study was set up in three provinces in 2018 and followed in six other provinces in 2020 and continued then again in ten more provinces in 2022. The programme is still going on, however, some figures are presented here. This is the first ever a large and nationally representative survey taking place face to face in households exclusively on the occurrence, causes, and management procedures for various types of disabilities and impairments in the country (full report will be published by SWO once becomes available).

The frequency distribution of the impairments in the country indicated that the movement (34%) and visual (20%) problems are of the highest rates of the impairments in Iran followed by mental, structure of brain and spinal cord, hearing, and voice and speech impairments (Table 12.4).

References

1. World Health Organization. Disability fact sheet. 2023. https://www.who.int/news-room/fact-sheets/detail/disability-and-health.
2. Statistical Centre of Iran. General population and housing census. 2006 and 2011. https://www.amar.org.ir.
3. Sajjadi H, Zanjari N. Disability in Iran: prevalence, characteristics, and socio-economic correlates. Arch Rehabil. 2015;16(1):36–47.
4. Moradi G, Mostafavi F, Hajizadeh M, Amerzade M, Mohammadi Bolbanabad A, Alinia C, Piroozi B. Socioeconomic inequalities in different types of disabilities in Iran. Iran J Public Health. 2018;47(3):427–34.
5. Rashidian A, Karimi-Shahanjarini A, Khosravi A, Elahi E, Beheshtian M, Shakibazadeh E, Khabiri R, Arab M, Zakeri M-R. Iran's multiple indicator demographic and health survey-2010: study protocol. Int J Prev Med. 2014;5(5):632.
6. Iranian Ministry of Health. Population and health in Iran. Tehran: Iranian Ministry of Health and Medical Education; 1992.
7. Iranian Ministry of Health. Health and diseases in Iran. Tehran: Iranian Ministry of Health and Medical Education; 2001.
8. Noorbala AA, Mohammad K. A summary of the National Health and Diseases Survey in Iran. Hakim. 2000;3(3):173–92.
9. State Welfare Organization of Iran. National Survey on the Prevalence and Causes of Disabilities and Impairments in Iran (RASAD). 2023. https://behzisti.rasad-disability.ir.

13 Disability in the TRoCA Region

Content

Tabriz Registry of Congenital Anomalies is located in the northwest of Iran as the details of the region are discussed in the second section of the book. Brief features of disability disorders are presented for this region separately in this chapter using the Iran national census of population and housing data for the years 2006 and 2011 [1].

The total prevalence rates based on the total number of disabilities in the regional population were estimated 21.10 (per 1000) for the year 2006, and 20.11 (per 1000) for 2011. The similar rates for the number of persons in the region who had at least one disability were calculated 18.59 (per 1000) and 11.62 (per 1000) for the years 2006 and 2011, respectively. Similarly to the national figures, the prevalence rates for males were higher than that of female rates (Table 13.1).

The overall data in the region indicated that nearly 17% of the individuals, either male or female, had two or more various types of disabilities at the same time.

Table 13.2 shows the prevalence rates of various types of disability between 2006 and 2011 in the region in the northwest of the country. Similarly to the national data, the prevalence rates of leg and foot defects and intellectual disabilities were the most common type of disability in this region. Males suffering from more types of disability compared to females in the TRoCA region. In addition to the intellectual, leg and foot defects, the voice and speech problems were also of the most frequent disabilities when the proportionally frequency distribution is taken into account in the regional population covered by the TRoCA programme (Table 13.3).

S. Dastgiri, *Congenital Anomalies in Iran*, https://doi.org/10.1007/978-981-97-6274-3_13

Table 13.1 Prevalence of disability in tabriz registry of congenital anomalies region (Iran National Census of Population and Housing Data)

	Prevalence (thousands)	Prevalence (thousands)
	Based on total number of disabilities	Based on individuals with at least one disability
Total (2006)	21.10	18.59
Male	27.26	24.25
Female	14.67	12.68
Total (2011)	20.11	11.69
Male	25.79	14.44
Female	14.27	8.93

Table 13.2 Prevalence of types of disabilities in tabriz registry of congenital anomalies region (Iran National Census of Population and Housing Data)

	Prevalence (thousands), 2006			Prevalence (thousands), 2011		
Disabilities	Total	Men	Female	Total	Male	Female
Visual	1.08	1.27	0.89	1.07	1.27	0.86
Hearing	1.19	1.32	1.05	1.26	1.40	1.11
Voice and speech	1.70	1.99	1.40	1.51	1.76	1.25
Arm and hand:						
Amputation	0.29	0.44	0.12	0.28	0.44	0.12
Defect	1.45	2.04	0.83	1.20	1.73	0.66
Leg and foot:						
Amputation	0.56	0.83	0.27	0.50	0.72	0.29
Defect	3.27	4.03	2.47	2.79	3.36	2.22
Trunk	1.36	1.61	1.10	1.37	1.60	1.14
Intellectual	3.94	4.89	2.95	4.09	4.95	3.21

Table 13.3 Frequency distribution of types of disabilities in tabriz registry of congenital anomalies region (Iran National Census of Population and Housing Data)

	Prevalence (thousands), 2006			Prevalence (thousands), 2011		
Disabilities	Total	Men	Female	Total	Male	Female
Visual	7.31	6.89	8.02	7.58	7.39	7.90
Hearing	8.01	7.15	9.49	8.93	8.11	10.26
Voice and speech	11.46	10.81	12.60	10.71	10.21	11.51
Arm and hand:						
Amputation	1.94	2.41	1.11	2.01	2.55	1.13
Defect	9.75	11.08	7.46	8.55	10.06	6.09
Leg and foot:						
Amputation	3.75	4.48	2.46	3.58	4.16	2.64
Defect	22.03	21.91	22.26	19.84	19.49	20.42
Trunk	9.17	8.73	9.95	9.74	9.29	10.46
Intellectual	26.58	26.54	26.65	29.06	28.74	29.59

Reference

1. Statistical Centre of Iran. General Population and Housing Census. 2006 and 2011. https://www.amar.org.ir.

Part IV

14 Background

The main aims of the community genetic programmes are to provide prevention, clinical, and psychosocial services for individuals and families suffering from congenital anomalies and hereditary conditions, and for those who are at risk of having those disorders later in the life. To achieve the goals of community genetic programmes, it is required that the following essential tasks are implemented in the community healthcare system: genetic consultation and intervention in a wide range of issues, accurate laboratory and clinical diagnosis of hereditary disorders and congenital anomalies, preconception care, prenatal and newborn screening for early childhood intervention, and timely and precise estimation of the hereditary risk in healthy individuals (carrier and population genetic screening).

An overview of the healthcare system in Iran will be described in this section. The definition, goals, and major characteristics of the community genetics services will then be discussed in the next chapter. Later, the primary, secondary, and tertiary level services and programmes will be reported that are currently available in Iran for the management and community control of congenital anomalies and genetic disorders.

S. Dastgiri, *Congenital Anomalies in Iran*, https://doi.org/10.1007/978-981-97-6274-3_14

How the Healthcare System Works in Iran? 15

Contents

The 29th article of the constitution of Iran indicates that every citizen in the country has the right to have the highest attainable level of health care [1]. The Ministry of Health and Medical Education is the main responsible entity in the country to fulfil this goal. The implementing bodies for the ministerial health policies are medical universities across the country. There are now 51 medical universities and 16 medical faculties running under the Ministry of Health and Medical Education. There is at least one public medical university in every province. There are also 96 medical schools working in private sector. The chancellor of a medical university is the highest official authority in each province working under the Health Minister. S/he is in charge of public health, care provision in public facilities, and medical education for the population in the province. Health care and public health services are provided through a nationwide network (Fig. 15.1). The network consists of a referral system beginning from primary care centres going through the secondary-level hospitals located mainly in the provincial capital, and hospitals in the tertiary level in major cities. The network gathers all health-related data from catchment areas making a nationwide information system.

The health system in each province includes a district health network. There is one district health centre in each district that administrating a number of urban and rural health centres in the region.

The network is well-established in the rural areas. Health houses are the first level of contact between the rural community and the network in the country. A health house, staffed by one or more health workers (called Behvarze), usually covers one or several nearby villages. Behvarzes are appointed by the Ministry of Health and Medical Education. They are recruited from the native individuals in the same community that the health house is located. Behvarze is well anchored in their community with long-standing close contact and relationship with individuals. Job description of Behvarzes includes annual census of the population covered, perinatal care, child care under 5 years, school health, family health services, immunization, case finding and referral, home visits for follow-up for elderly people.

Rural health centres are headed by physicians and staffed by paramedical and administrative

S. Dastgiri, *Congenital Anomalies in Iran*, https://doi.org/10.1007/978-981-97-6274-3_15

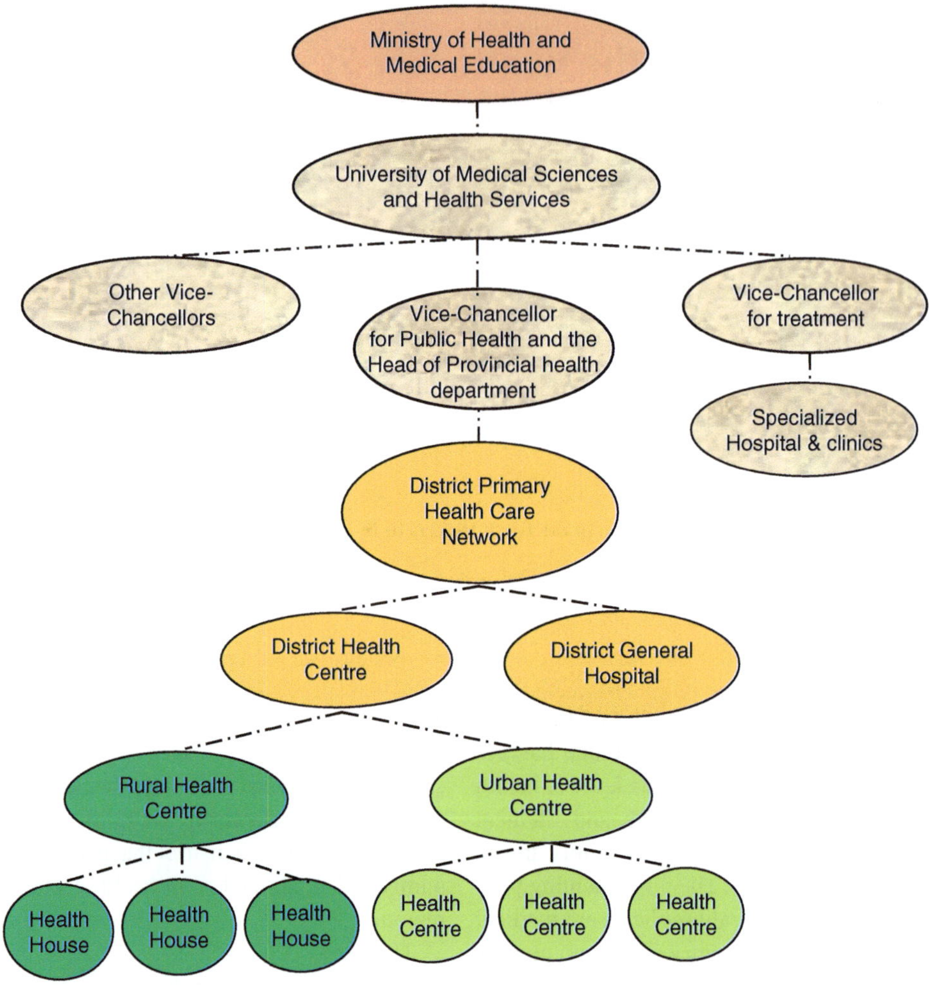

Fig. 15.1 Iranian primary health care system and health network

personnel. The team in the centre supervises the activities of the rural health houses, manages the referral cases from health houses, and maintains proper contact with both the higher levels of the health system and the district hospitals.

The public medical universities in provincial level provide primordial, primary, secondary, and tertiary services for the population under the coverage. They are almost the main provider of primary healthcare services across the country that are almost given free of charge in public facilities. Although a considerable part of the secondary and tertiary health services is provided by the medical universities in the country, they are widely available in private sectors as well [2–4].

Despite the outstanding achievements of the healthcare network in the country over the past few decades, mainly in deprived areas, the system seems now a little unfit for addressing the new demands and emerging needs, changes in the population dynamics and life span, the new shifts in disease epidemiological patterns towards chronic conditions, increasing burden of non-communicable diseases, migration of rural communities to urban areas, and growing expectations of the public for high-quality health and medical services.

15.1 Situation of the Birth Defects in Iran Healthcare System

As discussed in the earlier sections, the prevention and control programmes of congenital anomalies require that the accurate data are continuously collected and disseminated as part of the purposes for the surveillance system in the community. Using the surveillance programme, the public health authorities will be able to monitor the trends, to detect the emerging problems, to identify necessary interventions, and to determine the impact in a timely manner. The better the coverage and quality of the healthcare system, the more representative the information gathered through the surveillance programme.

Despite the availability of the evidence-based measures and best-practice guidelines for birth defects, the control and surveillance programmes often require a specific contextualization, and cross-cultural adaptation and validation in each region and country. In other words, data sources and quality of the surveillance programme might vary from one country to another depending on the socioeconomic and healthcare policy and practice, facilities in public and other related factors.

The performance of family physicians, as a major component in the healthcare system, has been found accurate enough in the diagnosis and clinical management of congenital anomalies. Healthcare system may consider family medicine as an effective approach to reduce the burden and high costs induced by medical technology and lab tests for the diagnosis of birth defects [5, 6].

Diagnosis and management of congenital anomalies are now included in the services provided by family physicians in the country. According to the protocols and programmes launched by the ministry of health, the family medicine team is supposed to screen and monitor all the newborns in terms of health and clinical status (including birth defects) in its specified catchment area. It seems that by effectively and efficiently implementing of the family medicine programme in the health system, there would be a unique opportunity to have a well-organized surveillance for monitoring of the congenital anomalies in the country [5–7]. The family medicine programme has been running since 2004 in most parts of the rural areas of Iran. There is, however, a long way to go for establishing a nationwide monitoring system integrated in the family medicine programme based on the best-practice guidelines for birth defects for the whole country in all urban and rural areas.

References

1. The 29th Article of the Constitution of Iran. 2023. https://ccri.ac.ir/fa/news/1707/%D8%A7%D8%B5%D9%84-29-%D9%82%D8%A7%D9%86%D9%88%D9%86-%D8%A7%D8%B3%D8%A7%D8%B3%DB%8C.
2. Iranian health houses open the door to primary care. Bull World Health Organ. 2008;86(8):585–6.
3. Mehrdad R. Health system in Iran. JMAJ. 2009;52(1):69–73.
4. Shadpour K. Primary health care networks in the Islamic Republic of Iran. East Mediterr Health J. 2000;6(4):822–5.
5. Mashhadi Abdolahi H, Kargar Maher MH, Karamouz M, Afsharnia F, Khosroshahi H, Dastgiri S. How accurate is diagnosis of congenital anomalies made by family physicians? Health Promot Perspect. 2014;4(2):158–64.
6. Mashhadi Abdolahi H, Kargar Maher MH, Karamouz M, Khosroshahi H, Dastgiri S. Reliability and validity of the March of Dimes preconception/prenatal family health history questionnaire: the Persian version. Congenit Anom (Kyoto). 2016;56(3):107–11.
7. Takian AH, Rashidian A, Kabir MJ. Expediency and coincidence in re-engineering a health system: an interpretive approach to formation of family medicine in Iran. Health Policy Plan. 2011;26:163–73.

16 What Are Community Genetics Services?

Content

The occurrence rates of congenital and genetic disorders in low- and middle-income countries seem to be higher because of a number of reasons including low availability of community genetics services, the higher consanguinity rates, large family size, insufficient of laws to support termination of pregnancy, and advanced maternal age at pregnancy. However, compared to the control of communicable diseases and non-communicable chronic disorders, public health genetics services are not among the main priorities of health system in those areas due to the present importance and ongoing burden of communicable diseases in the community, lack of the organized and functional public agencies for the treatment and prevention of specific genetic diseases, lower resources (i.e. genetic laboratory and technology, funding, trained professionals, and manpower), lower genetics literacy, paucity of epidemiology of genetic and congenital disorders, and cultural, legal, societal, and religious limitations [1, 2].

Community genetics is defined as "the art and science of the responsible and realistic application of health and disease-related genetics and genomics knowledge and technologies in human populations and communities to the benefit of individuals therein. Community genetics is multi-, inter- and transdisciplinary and aims to maximize benefits while minimizing the risk of harm, respecting the autonomy of individuals and ensuring equity". The core activities in community genetics include screening, consultation and education, quality control of genetic services, genetics in primary health care and prevention, genetics in population subgroups (i.e. ethnic minorities), registries, preconception care, epidemiology, policy, societal, ethical, and legal issues [3]. In summary, the principal objectives of the community genetic services are to manage the care activities in the area of clinical, preventive, and psychosocial demands of those suffering from/at risk of genetic diseases and congenital anomalies by planning and implementing the precise and timely diagnosis and healthcare programmes.

Congenital and genetic disorders, as a component of maternal and child health, has been accepted as one of five public health priorities within the framework of health system strengthening among the WHO Eastern Mediterranean Region countries. However, although the main concepts and approaches of community genetics is basically a nascent field in the LMIC world, some countries in the WHO Eastern

S. Dastgiri, *Congenital Anomalies in Iran*, https://doi.org/10.1007/978-981-97-6274-3_16

Mediterranean Region (mainly in Bahrain and Iran) have successfully implemented control and prevention programmes for a number of congenital and genetic disorders [4, 5].

For planning and developing the genetics services at the community level, the following essentials are required as recommended by the world health organization experts group of specialists in the field of community genetics:

- Commitment of policy makers: this commitment would broadly be required for prioritizing the congenital and genetic disorders as one of the basic primary health care components, funding and managing the community level prevention and control programmes.
- Integration of community genetics services into the existing primary health system and reproductive health centres: the integration includes the current preconception and post-conception counselling and care programmes, risk detection in couples, carrier screening, newborn screening, and care for the affected individuals and their high-risk relatives.
- Setting up the registries for selected conditions and groups of congenital and genetic disorders: the registry may initially be set up in a local, regional, or national level depending on the available funds and resources.
- Population screening programmes: this includes newborn screening (i.e. phenylketonuria, congenital hypothyroidism, genetic blood disorders and cystic fibrosis), carrier screening in healthy individuals (i.e. thalassaemia), and prenatal screening (i.e., chromosome abnormalities, haemoglobinopathies, neural tube defects, and other foetal congenital anomalies).
- Establishing genetics centres equipped with new molecular and advanced technologies.
- Care programmes for affected individuals with genetics and congenital disorders: the programmes include, for instance, surgical correction for some congenital malformations, dietary support and treatment for some metabolic diseases and blood disorders, and rehabilitation for some types of mental and physical disabilities [2, 4].

- *training health professionals in basic concepts of genetics and their application to community genetics services.*
- *use of family history as an instrument to detect genetic risks.*
- *preconception and prenatal care, including folic acid fortification or supplementation, addressing the genetic risks of maternal age and consanguinity, genetic counselling when needed, prenatal ultrasound screening to detect fetal abnormalities, and referring detected genetic problems to a tertiary Centre for proper management.*
- *public education to avoid alcohol, tobacco and potential teratogens in gestation.*
- *carrier detection for haemoglobinopathies in countries where these conditions are a significant burden to public health, followed by genetic counselling and the option of prenatal diagnosis where voluntary termination of affected pregnancies is acceptable.*
- *clinical detection of congenital disorders in newborns and appropriate referrals to higher levels of care for diagnosis and treatment.*
- *newborn screening for congenital conditions in which early intervention is effective, such as congenital hypothyroidism, followed by treatment and other services for life. Newborn screening with proper management for genetic conditions common in each country could also be considered.*

In principle, the management of congenital and genetic disorders, as for the most diseases in the human population, are performed at five levels of prevention including primordial (i.e. decreasing the risk factors exist in the community), primary (i.e. genetic consultation), secondary (i.e. drug treatments for thyroid anomalies), tertiary (i.e. rehabilitation for limb defects), and quaternary (i.e. avoiding over-medicalization, unnecessary and excessive intervention).

References

1. March of Dimes. Global report on birth defects: the hidden toll of dying and disabled children. New York: March of Dimes Birth Defects Foundation; 2006. http://www.marchofdimes.org/materials/global-report-on-birth-defects-the-hiddentoll-of-dying-and-disabled-children-full-report.pdf.
2. World Health Organization. Community genetics services: report of a WHO consultation on community genetics. Geneva: World Health Organization; 2011.
3. Ten Kate LP, Al-Gazali L, Anand S, Bittles A, Cassiman JJ, Christianson A, Cornel MC, Hamamy H, Kääriäinen H, Kristoffersson U, Marais D, Penchaszadeh VB, Rahman P, Schmidtke J. Community genetics. Its definition 2010. J Community Genet. 2010;1(1):19–22.
4. World Health Organization. Shaping the future of health in the WHO eastern Mediterranean region: reinforcing the role of WHO. Cairo: WHO Regional Office for the Eastern Mediterranean; 2012. http://applications.emro.who.int/dsaf/EMROPUB_2012_EN_742.pdf.
5. World Health Organization. Expert meeting on the prevention of congenital and genetic disorders in the Eastern Mediterranean Region, London, United Kingdom of Great Britain and Northern Ireland 29–31 July 2016. http://applications.emro.who.int/docs/IC_Meet_Rep_2016_EN_18989.pdf?ua=1.

Community Genetics Services in Iran

17

Contents

17.1 Background

As discussed earlier, due to the epidemiological transition caused by the success of programmes in the prevention and control of infectious diseases and the health of children and mothers over the past few decades, the burden of genetic and congenital disorders has proportionally increased in some of the low- and middle-income countries. In addition, although the hereditary diseases are mostly rare and the contribution of genetic factors are few in each disease, the overall contribution and total prevalence figures for the genetic and congenital disorders are considerably high. Development and management of community genetics programmes for the control of occurrence and prevention of congenital disorders and genetic diseases should, therefore, be among the priorities of the health system in those areas.

Community genetics services in Iran have been recognized, as a priority, in the health system since two decades ago. Prevention and control programmes for genetic disorders and congenital anomalies are then developed and implemented into the country's health system based on the increasing occurrence figures, remarkable changes in the burden and epidemiological pattern of congenital and

S. Dastgiri, *Congenital Anomalies in Iran*, https://doi.org/10.1007/978-981-97-6274-3_17

genetic disorders in the country in the recent years [1–5].

In the absence of preventive interventions, it is expected that for every thousand live births in the country, nearly 48 cases of genetic and hereditary disorders occur [1]. Some other estimates indicate that more than 70 cases of congenital and genetic abnormalities occur annually in every thousand live births in Iran [6]. However, in both cases, if the community genetics services are implemented, these figures can decrease by more than 70% [1].

17.2 Care Services Available for the Community Genetics in Iran

The current care services for the prevention, clinical management, and rehabilitation of genetic and congenital disorders available in the public and private sectors in Iran are briefly reported here [7].

17.2.1 Primordial and Primary Prevention

- Public genomics educational programmes
- Preconceptional care including folic acid and multivitamin supplementation
- Healthcare services during the pregnancy
- Maternal and children health services including vaccination for rubella and measles

17.2.2 Screening for Selected Disorders

- Congenital hypothyroidism
- Thalassaemia
- Neural tube defects
- Down syndrome
- Phenylketonuria

The details of the screening programmes that are currently carried out for those anomalies in the country will be discussed later in this chapter.

17.2.3 Diagnosis Procedures Available for Genetic and Congenital Defects

- Ultrasonography
- DNA extract and molecular tests
- Serum alpha foetoprotein
- Nonconjugated oestradiol
- Amniocentesis
- Chromosomal analysis of amniotic fluid
- Maternal phenylalanine test
- Plasma amino acids analysis foetal echocardiography
- Magnetic Resonance Imaging (MRI) scan and Computed Tomography (CT) scan
- Electroencephalography
- Chest radiography
- Electrocardiogram and echocardiography
- Arteriography
- Aortography
- Colour ultrasonography
- Pulse and colour Doppler
- Angiocardiography
- Pulmonary angiography
- Liver function tests
- Intravenous urography and urethrography
- Blood gas analysis
- Chest radiography
- Radiography by barium swallow
- Thoracentesis
- Bronchography
- Lung biopsy and lung aspiration
- Bronchoscopy, laryngoscopy, thoracoscopy
- Radionucleotide scan
- Fluoroscopy

17.2.4 Treatment and Clinical Management

Various types of drug treatment, surgical correction, and rehabilitation procedures are routinely available both in private and public sectors in the country mostly for major defects and congenital malformations (i.e. heart defects, hydrocephaly and spina bifida, cleft lip and palate, urogenital disorders).

It must be noted that genetic services, diagnosis, treatment, and rehabilitation approaches are usually expensive facilities for congenital and genetic disorders. Despite they are routinely available in private sector in Iran, most people do not afford to get access to those services if needed. Necessary resources have to, therefore, be allocated for public sector and healthcare system in the country to make the essentials of community genetic services easily available for affected individuals and families who need.

17.3 Pillars in the Current Guideline of Community Genetics Services in Iran

Once the importance and necessity of community genetics services was recognized by the public health policy making bodies in Iran, a comprehensive guideline for community genetics services was introduced for implementing in the healthcare system by the Iran ministry of health and medical education in 2007. The programme was updated in 2022 based on the new findings and successes in the area of genetic services in the country [8]. The core principles of this guideline are as follows:

17.3.1 Principles of Genetics Services Implementation in the Community

- Genomics education
- Screening and detection of high/at risk individuals and families
- Genetic diagnosis
- Genetic counselling (before and after genetic diagnosis)
- Genetic services at primary healthcare system and risk reduction programmes

17.3.2 Job Description of the Community Genetics Bureau

- Developing of community genetics programmes and coordinating with stakeholders
- Heading the national advisory committee and technical sub-committees in the field of community genetics including training of care providers, identification of individuals and families needing genetic facilities, and care services (i.e. diagnosis, counselling)

17.3.3 Checklists, Forms, and Data Management

A wide variety of forms and checklists are routinely filled for clients in the healthcare system across the country. Using the predesigned forms, the data are gathered in a central database for genetic services and screening for couples for marriage-related matters, regular occurrence report from hospitals and clinics (both in public and private sectors) for selected groups of genetic and congenital disorders, and prenatal and neonatal screening for metabolic hereditary diseases, etc. The data are analysed for epidemiological, monitoring, and surveillance purposes as planned in the guideline.

17.3.4 Monitoring and Evaluation of the Programmes

The aim of monitoring and evaluation of the programmes implemented by the genetics bureau in the country is to assess whether the objectives are fulfilled and whether the activities are right on the way according to the guideline, and as it has been planned. Monitoring and evaluation are performed on annual basis by the genetics bureau using the predefined indices for each programme.

17.3.5 National Screening Programmes for the Prevention of Selected Genetic and Congenital Disorders

The screening for the prevention of beta thalassaemia started in early 1990s in some provinces in Iran. Later in 1997, the national screening programme was formally integrated in the country's health system.

Since then and after outstanding achievements of the thalassaemia programme, it has been broadly accepted as a successful model for integration of the genetic services in the country's health system for some other congenital and genetic disorders with specific procedures for each condition including:

- Down syndrome
- Phenylketonuria
- Congenital hypothyroidism
- Haemoglobinopathies
- Hereditary metabolic diseases
- Hereditary bleeding diseases
- Hereditary neuromuscular diseases
- Hereditary deafness
- Hereditary blindness
- Hereditary immune deficiency

In addition to the above conditions, familial aggregation for hereditary diseases is a matter of concern in the national screening programmes for prevention. Accordingly, the presence of a positive family history for a particular disease and repeating of a disease in relatives (i.e. the presence of at least two patients suffering from the same disease) would be considered as the risk of familial aggregation or carriers for hereditary diseases in the guideline. Thus, if healthcare provider classified a person as having the "probability of genetic risk factor" or "likely to suffer from a hereditary disease", a non-urgent referral to a genetic service is made.

17.4 Successes Examples in the Community Genetics Services in Iran

Nearly 7.9 million infants are born every year with birth defects in the world, the majority (7.4 million) in the low- and middle-income countries with 43% of case fatality rate among under age 5 year. The blood hereditary disorders have a significant contribution to the global burden due to the total birth defects. It is estimated that 7% of the world population are carriers for haemoglobin disorders [9]. Those blood disorders including alpha and beta thalassaemia mostly occur in the tropical communities in the sub-Saharan Africa, Mediterranean, Middle East, and southeast Asian countries. In the eastern Saudi Arabia, approximately half of the people have an asymptomatic form of alpha thalassaemia. These regions are located in the area of high occurrence that is usually called the "thalassaemia belt" in the globe. The overall carrier rate for beta thalassaemia is estimated between 1% and 20% in different communities in the world [10–13]. In thalassaemia, as one of the most common genetic diseases, the production of normal haemoglobin is totally or partially impaired. The complications of the beta thalassaemia disease include enlargement of the liver, spleen, and bones, thinning of the bones and risk of fracture, susceptibility to infection, growth problems, and iron complications in the heart, pancreas, and liver, resulting in diabetes, heart, and liver failure in patient.

Beta thalassaemia major is the most common hereditary blood disease in Iran. The carrier rate of thalassaemia gene in the country is about 4%, which is remarkably higher than that of the average figure in the world (1.5%). The carrier rate of thalassaemia gene is not the same in different parts of the country. In the coastal areas, in the north and the Persian Gulf and the Oman Sea in the southern regions, it is more common than other regions. Estimates show that the provinces of Khuzestan, Bushehr, Hormozgan, Sistan and

Baluchestan, Kerman, Gilan, Mazandaran, and even in the provinces of the central regions such as Isfahan and Fars up to 10% of the population might carry the gene of this disease [8].

After careful evaluation of the pilot studies in some provinces in Iran between 1990 and 1996, the nationwide screening and prevention programme of thalassaemia formally started in Iran in 1996–7 aiming to detect healthy carriers in couples before they are formally married. Once a positive case identified in the country's health network, the couple is offered various counselling and providing them with the opportunity either to separate or prenatal diagnosis, if pregnant, and selective termination of pregnancy in the future before the 16th weeks of gestation for thalassaemia. According to the country's law and regulations, the cases diagnosed at the risk of thalassaemia has to be legalized and approved for abortion by a team of religious verdict and legal medicine specialists. The ethical and social sensitivities and values in Iran are carefully taken into account in every part of the programme. For instance, to avoid not to stigmatize a woman with a possible risk of genetic disorder, once a man was initially tested as a positive carrier of thalassaemia, it would then be for his female partner to take the screening test later [11–16].

After implementing the prevention and control programme of thalassaemia, the incidence rate of disease has decreased from 1 case per 1000 live births in 1996 (before the start of the programme) to nearly 1 case per 10,000 births in 2019 indicating that this prevention programme of thalassaemia in Iran has played an effective role in reducing the thalassaemia birth occurrence. This is a remarkable achievement for a vast, and middle-income country with a population of 85 million. The overall success rate for the screening and prevention of thalassaemia in the country is estimated in every year of the programme in each province using the formula: [1 − (observed thalassaemia new cases/estimated expectation of thalassaemia new cases)] × 100 [8].

The Iranian prevention and control programme of thalassaemia might be useful for countries of "thalassaemia belt", and for countries with ethnic minority communities (in the areas of high prevalence) where the political, geographical, social, and economic status may even be similar or very different.

Besides this great achievement in the control of thalassaemia births in the country, it is still for policy makers in Iran to upgrade the ongoing strategies and approaches helping the health system to remove the current obstacles and barriers for further reduction of the number of the thalassaemia-affected patients in the Iranian population.

References

1. Azimi SS, Samavat SA, Rezaei Tabar V, Soori H. Effectiveness of the preventive intervention of chromosomal disorders of Iran's community genetics program: application of Bayesian network. Med J Islam Repub Iran. 2021;35:132.
2. Behnam B, Zakeri M. Genetics and genomic medicine in Iran. Mol Genet Genomic Med. 2019;7(2):e00606.
3. Atri Barzanjeh S, Behshid M, Hosseini MB, Ezari M, Taghizadeh M, Dastgiri S. Community genetic services in Iran. Genet Res Int. 2012;2012:129575.
4. Christianson A, Streetly A, Darr A. Lessons from thalassaemia screening in Iran. BMJ. 2004;329(7475):1115–7.
5. Aghajani H, Samavat A, Haghazali M, Valizadeh F, Sarbazi G. Primary health care: an approach to community control of genetic and congenital disorders. Iran J Public Health. 2009;38(Suppl 1):113.
6. Christianson A, Howson CP, Modell B. March of Dimes global report on birth defects: the hidden toll of dying and disabled children. March of Dimes birth defects foundation; 2006. https://dev.marchofdimes.org/materials/global-report-on-birth-defects-the-hidden-toll-of%2D%2Dd2unzZI5_VWOaLZnw6iHcx7hbpMWtWzTuIOU3DabcVY.pdf.
7. Ezari M. Community Genetic Services in Iran. Postgraduate Thesis Submitted to the Tabriz University of Medical Sciences; 2011. (note: author and his team in Tabriz Registry of Congenital Anomalies (TRoCA) updated the data for this chapter in 2023 for the purpose of this book using the information gathered through the face-to-face interviews,

meetings and visits at the selected community genetics facilities.)
8. Iran Ministry of Health. Community Genetics Bureau. Guideline of Community Genetics Services in Iran Health System. Tehran: Ministry of Health; 2022.
9. World Health Organization. Management of Birth Defects and Haemoglobin Disorders: Report of a Joint WHO-March of Dimes Meeting. Geneva: World Health Organization; 2006. https://iris.who.int/bitstream/handle/10665/43587/9789241594929_eng.pdf?sequence=1.
10. Beta thalassemia. Wikipedia; 2023. https://en.wikipedia.org/wiki/Beta_thalassemia.
11. Dastgiri S. Epidemiology of congenital anomalies and genetic disorders. In: Yavari P, editor. Epidemiology of prevalent diseases in Iran. GAP Publications; 2014. (Persian).
12. Samavat A, Modell B. Iranian national thalassaemia screening programme. BMJ. 2004;329(7475):1134–7.
13. Najmabadi H, Ghamari A, Sahebjam F, Kariminejad R, Hadavi V, Khatibi T, et al. Fourteen-year experience of prenatal diagnosis of thalassemia in Iran. Community Genet. 2006;9(2):93–7.
14. Abolghasemi H, Amid A, Zeinali S, Radfar MH, Eshghi P, Rahiminejad MS, et al. Thalassemia in Iran: epidemiology, prevention, and management. J Pediatr Hematol Oncol. 2007;29(4):233–8.
15. Strauss BS. Genetic counseling for thalassemia in the Islamic Republic of Iran. Perspect Biol Med. 2009;52(3):364–76.
16. Miri M, Tabrizi Namini M, Hadipour Dehshal M, Sadeghian Varnosfaderani F, Ahmadvand A, Yousefi Darestani S, et al. Thalassemia in Iran in last twenty years: the carrier rates and the births trend. Iranian J Blood Cancer. 2013;6(1):11–7.

Part V

18 Concluding Remarks

Contents

18.1 Registry Programme for Congenital Anomalies

Occurrence of genetic disorders and congenital anomalies are now making a proportionally greater contribution to perinatal mortality and ill health in childhood in many places in the world both in the most developed areas and in low- and middle-income countries. Monitoring and surveillance systems for the cases of genetic and birth defects are now being performed in more than 50 communities across the globe, mostly located in the high-income and developed countries.

This book reports the activities of the Tabriz Registry of Congenital Anomalies (TRoCA) aiming to provide a community-based overview of the congenital anomalies, disabilities, and community genetics services in Iran as one of the low- and middle-income countries:

- In the year 2000, TRoCA was established in the northwest of Iran, as an exclusive programme for monitoring the occurrence of congenital anomalies in the population. The defect cases in the TRoCA registry comprised all births with a primary diagnosis of congenital anomalies under one of the main headings defined in the International Classification of Diseases (ICD). TRoCA is a member of the International Clearinghouse for Birth Defects Surveillance Research (ICBDSR) and a number of other global entities working in the field of congenital anomalies. The overall implication of the TRoCA is to determine the contribution of the congenital anomalies occurrence to the whole community health in the population under coverage in TRoCA. This would give an estimate of the size of the health problem raised by genetic and congenital anomalies in the population. The programme would ultimately aim to implement control plans for the prevention of genetic disorders and congenital anomalies in the community.
- TRoCA presents the baseline data of congenital anomalies in the region. It also documents the changes of the occurrence of anomalies over time. This may be useful in identifying

S. Dastgiri, *Congenital Anomalies in Iran*, https://doi.org/10.1007/978-981-97-6274-3_18

clues to the aetiology of some defects including genetic and environmental factors. These factors are preventable by proper intervention in early stage of occurrence. Prevalence data might, too, have an important role in planning and evaluating the strategies for primary prevention of congenital anomalies, particularly in high-risk groups.
- Implementation of preventive measures is not possible without having access to adequate and valid sources of data. To be able to do that, in addition to the public health sector, the related disciplines including various hospital services, and specialist clinical genetics and foetal medicine departments should also be involved. This would eventually provide specialists in various disciplines a rare source of data to assess the mortality, abortion, and factors influencing the prognosis of anomalies in various age groups. This information may also be important for families in planning healthcare needs for their affected children.
- An appropriate decision on the therapeutic termination of pregnancy depends on professionals and parents to have access to proper valid information. This would particularly be important for the lethal congenital defects (i.e. respiratory system anomalies, neural tube defects, and other nervous system anomalies).
- This unique source of information in TRoCA is essential for public health workers in planning healthcare services for children born with birth defects, and for education professionals in planning special schools and educational materials for handicapped children affected by congenital anomalies.
- The TRoCA data about causes of death in congenital anomalies may be useful in identifying preventable causes of death in different groups of anomalies. This will help parents and health professionals to prevent avoidable early deaths in children affected by anomalies.
- The management, organizational capacity, and regional and international achievements of the TRoCA programme may be specific to the study population whilst the core structure of the programme may have more general applications in similar areas, and for those professional in low- and middle-income communities willing to set up a similar facility for monitoring the occurrence of congenital anomalies.

18.2 Control of Congenital Anomalies

The control and care principles of genetic and congenital diseases for implementing in the healthcare system include six major cores:

- Preconceptional and pregnancy health care (i.e. screening for congenital and genetic disorders, quitting smoking and certain medications, taking folic acid and multivitamin supplementations)
- Genetic counselling (i.e. avoiding the marriage for some heterozygous diseases, avoiding pregnancy, or termination of pregnancy for some disorders)
- Genetic screening (i.e. detection of healthy carriers, early detection of genetic diseases using the easy to access clinical and laboratory technologies)
- Monitoring of the population (i.e. epidemiology, inheritance pattern, and estimating the occurrence for certain congenital and genetic disorders in the community)
- Genetic education (i.e. planning for the active participation of social groups, families, and non-governmental organizations in the control and prevention programmes of genetic diseases)
- For a substantial success in the control of genetic disorders at the community level, besides all the existing prevention strategies, a particular attention has to always be paid on interventions before conception and embryo formation.

18.3 The National Prevention and Control Programmes in Iran

Ministry of Health in Iran has currently integrated the following programmes in the national healthcare system for prevention and control of the occurrence of genetic and congenital disorders in the country:

- Prenatal screening for trisomies and abnormalities
- Screening for genetic disorders (i.e. thalassaemia carriers) in couples before marriage
- Newborn screening for inherited metabolic diseases
- Family history and genetic risk assessment for certain chronic diseases (i.e. breast and colon cancer, and coronary heart diseases in the youth and middle-aged individuals)
- Genetic counselling in a wide range of choices (including therapeutic termination of pregnancy), if any of those conditions identified

Diagnosis and clinical management, surgical procedures, drugs, and physiotherapy for genetic disorders are usually costly making the affected individuals and families need special attention, further aids and financial support. Some non-governmental organizations and non-for-profit societies and associations (i.e. Rare Diseases Foundation, Multiple Sclerosis Association, Haemophilia Association, Autism Society) are now closely working along with the activities of ministry of health helping those individuals and families suffering from congenital and genetic diseases in a wide range of issues.

18.4 Conclusions

Efforts to reduce the burden due to the birth defects have consisted of various levels and approaches for prevention. Noticeable progress has already been made through primary prevention efforts with the preconceptional folic acid supplementation. This may contribute to the long-term reduction of the occurrence of congenital anomalies if compliance to health professionals' recommendations on this matter is fully observed by women. However, we need not forget that until it becomes possible to reduce birth defects rate by the primordial or primary prevention approaches more efficiently, secondary and tertiary strategies of prevention are needed to minimize suffering caused by congenital anomalies and to promote the patients' adjustment to their existing disorder. Furthermore, some targeted educational and health service interventions directed at children with congenital anomalies are needed to reduce further the probability of death and increase their life expectancy.

There are always barriers and unmet needs due to the demographic and cultural characteristics in Iran (similar to other low- and middle-income communities), and growing advancements and ever-increasing information in the field of genetics around the world. Actually, we are in the beginning of a long way ahead to fulfil the objectives of the TRoCA programme. More works, for instance, are required in prevention plans, and to assess the quality of life as well as survival pattern in children with defects so that their healthcare needs may be better understood and met.

The financial sanctions and political tensions over the past several years have made it very difficult, even impossible, to apply for new funds and grants to continue with the objectives of the TRoCA programme for registration and prevention of congenital anomalies. Despite those barriers and challenges, everyone in our team believes that we should, for the sake of people's health, stay firm and keep the light of TRoCA alive on the road of science and health!

Appendices

Appendix A: ICD-11 for Mortality and Morbidity Statistics (Version : 01/2023)

20 Developmental anomalies
Structural developmental anomalies primarily affecting one body system
Structural developmental anomalies of the nervous system
LA00 Anencephaly or similar anomalies
LA00.0 Anencephaly
LA00.00 Craniorachischisis
LA00.0Y Other specified anencephaly
LA00.0Z Anencephaly, unspecified
LA00.1 Iniencephaly
LA00.2 Acephaly
LA00.3 Amyelencephaly
LA00.Y Other specified anencephaly or similar anomalies
LA00.Z Anencephaly or similar anomalies, unspecified
LA01 Cephalocele
LA02 Spina bifida
LA02.0 Spina bifida cystica
LA02.00 Myelomeningocele with hydrocephalus
LA02.01 Myelomeningocele without hydrocephalus
LA02.02 Myelocystocele
LA02.0Y Other specified spina bifida cystica
LA02.0Z Spina bifida cystica, unspecified
LA02.1 Spina bifida aperta
LA02.Y Other specified spina bifida
LA02.Z Spina bifida, unspecified
LA03 Arnold-Chiari malformation type II
LA04 Congenital hydrocephalus
LA04.0 Hydrocephalus with stenosis of the aqueduct of Sylvius
LA04.Y Other specified congenital hydrocephalus
LA04.Z Congenital hydrocephalus, unspecified
LA05 Cerebral structural developmental anomalies
LA05.0 Microcephaly
LA05.1 Megalencephaly
LA05.2 Holoprosencephaly
LA05.3 Corpus callosum agenesis
LA05.4 Arrhinencephaly

S. Dastgiri, *Congenital Anomalies in Iran*, https://doi.org/10.1007/978-981-97-6274-3

20 Developmental anomalies
LA05.5 Abnormal neuronal migration
LA05.50 Polymicrogyria
LA05.51 Cortical dysplasia
LA05.5Y Other specified abnormal neuronal migration
LA05.5Z Abnormal neuronal migration, unspecified
LA05.6 Encephaloclastic disorders
LA05.60 Porencephaly
LA05.61 Schizencephaly
LA05.62 Hydranencephaly
LA05.6Y Other specified encephaloclastic disorders
LA05.6Z Encephaloclastic disorders, unspecified
LA05.7 Brain cystic malformations
8D67 Intracranial arachnoid cyst
LA05.8 Colpocephaly
LA05.Y Other specified cerebral structural developmental anomalies
LA05.Z Cerebral structural developmental anomalies, unspecified
LA06 Cerebellar structural developmental anomalies
LA06.0 Dandy-Walker malformation
LA06.1 Hypoplasia or agenesis of cerebellar hemispheres
LA06.2 Focal cerebellar dysplasia
LA06.Y Other specified cerebellar structural developmental anomalies
LA06.Z Cerebellar structural developmental anomalies, unspecified
LA07 Structural developmental anomalies of the neurenteric canal, spinal cord or vertebral column
LA07.0 Primary tethered cord syndrome
LA07.1 Diastematomyelia
LA07.2 Amyelia
LA07.3 Primary syringomyelia or hydromyelia
LA07.4 Arnold-Chiari malformation type I
LB73.0 Occult spinal dysraphism
LA07.Y Other specified structural developmental anomalies of the neurenteric canal, spinal cord or vertebral column
LA07.Z Structural developmental anomalies of the neurenteric canal, spinal cord or vertebral column, unspecified
LA0Y Other specified structural developmental anomalies of the nervous system
LA0Z Structural developmental anomalies of the nervous system, unspecified
Structural developmental anomalies of the eye, eyelid or lacrimal apparatus
LA10 Structural developmental anomalies of ocular globes
LA10.0 Microphthalmos
LA10.1 Clinical anophthalmos
LA10.2 Buphthalmos
LA10.3 Congenital macrophthalmos
LA10.Y Other specified structural developmental anomalies of ocular globes
LA10.Z Structural developmental anomalies of ocular globes, unspecified
LA11 Structural developmental anomalies of the anterior segment of eye
LA11.0 Blue sclera
LA11.1 Structural developmental anomalies of cornea
9A78.51 Corneal staphyloma
LA11.2 Anterior segment dysgenesis
LA11.3 Aniridia
LA11.4 Coloboma of iris
LA11.5 Congenital corneal opacity

20 Developmental anomalies
LA11.6 Structural disorders of the pupil
LA11.60 Irregular pupil of the eye
LA11.61 Iridoschisis
LA11.62 Anomalies of pupillary function
LA11.6Y Other specified structural disorders of the pupil
LA11.6Z Structural disorders of the pupil, unspecified
9C61.4 Developmental glaucoma
9C61.40 Primary congenital glaucoma
9C61.41 Primary infantile glaucoma
9C61.42 Secondary childhood glaucoma
LA11.3 Aniridia
LD28.01 Marfan syndrome
1F02 Rubella
1F02.0 Rubella with neurological complications
1F02.1 Rubella arthritis
1F02.2 Rubella without complication
KA62.8 Congenital rubella syndrome
1F02.Y Rubella with other specified complication
5C60.0 Oculocerebrorenal syndrome
LD2D.1 Neurofibromatoses
LD2D.10 Neurofibromatosis type 1
LD2D.11 Neurofibromatosis type 2
LD2D.12 Neurofibromatosis type 3
LD2D.1Y Other specified neurofibromatoses
LD2D.1Z Neurofibromatosis, unspecified
9C61.4Y Other specified developmental glaucoma
9C61.4Z Developmental glaucoma, unspecified
LA11.Y Other specified structural developmental anomalies of the anterior segment of eye
LA11.Z Structural developmental anomalies of the anterior segment of eye, unspecified
LA12 Structural developmental anomalies of lens or zonula
LA12.0 Coloboma of lens
LA12.1 Congenital cataract
LA12.2 Congenital aphakia
LA12.3 Spherophakia
LA12.Y Other specified structural developmental anomalies of lens or zonula
LA12.Z Structural developmental anomalies of lens or zonula, unspecified
LA13 Structural developmental anomalies of the posterior segment of eye
LA13.0 Congenital anomalies of the vitreous
LA13.3 Congenital vitreoretinal dysplasia
LD27.00 Incontinentia pigmenti
LA13.1 Coloboma of choroid or retina
LA13.2 Coloboma of macula
LA13.3 Congenital vitreoretinal dysplasia
LD27.00 Incontinentia pigmenti
LA13.5 Congenital retinal aneurysm
LA13.6 Congenital malformations of choroid
LA13.7 Congenital malformation of optic disc
LA13.70 Isolated optic nerve hypoplasia
LA13.71 Optic nerve aplasia
LA13.72 Congenitally elevated optic disc
LA13.73 Optic disc dysplasia

20 Developmental anomalies
LA13.74 Megalopapilla
LA13.76 Coloboma of optic disc
LA13.7Y Other specified congenital malformation of optic disc
LA13.7Z Congenital malformation of optic disc, unspecified
LA13.8 Certain congenital malformations of posterior segment of eye
LA13.80 Anastomosis of retinal or choroidal vessels
LA13.1 Coloboma of choroid or retina
9B73.11 Juvenile retinoschisis
LA13.Y Other specified structural developmental anomalies of the posterior segment of eye
LA13.Z Structural developmental anomalies of the posterior segment of eye, unspecified
LA14 Structural developmental anomalies of eyelid, lacrimal apparatus or orbit
LA14.0 Structural developmental anomalies of eyelids
LA14.00 Palpebral cleft or coloboma
LA14.01 Cryptophthalmia
LA14.02 Congenital entropion
LA14.03 Congenital ectropion
LA14.04 Congenital ptosis
LA14.05 Congenital eyelid retraction
LA14.06 Epibulbar choristoma
LA14.07 Ankyloblepharon filiforme adnatum
LA14.0Y Other specified structural developmental anomalies of eyelids
LA14.1 Structural developmental anomalies of lacrimal apparatus
LA14.10 Aplasia of lacrimal or salivary glands
LA14.11 Agenesis of lacrimal ducts
LA14.12 Congenital dacryocele
LA14.13 Congenital agenesis of lacrimal punctum
LA14.14 Congenital stenosis or stricture of lacrimal duct
LA14.1Y Other specified structural developmental anomalies of lacrimal apparatus
LA14.1Z Structural developmental anomalies of lacrimal apparatus, unspecified
LA14.2 Structural developmental anomalies of orbit
LA14.Y Other specified structural developmental anomalies of eyelid, lacrimal apparatus or orbit
LA14.Z Structural developmental anomalies of eyelid, lacrimal apparatus or orbit, unspecified
LA1Y Other specified structural developmental anomalies of the eye, eyelid or lacrimal apparatus
LA1Z Structural developmental anomalies of the eye, eyelid or lacrimal apparatus, unspecified
Structural developmental anomalies of the ear
LA20 Structural anomaly of eustachian apparatus
LA21 Minor anomalies of pinnae
LA21.0 Macrotia
LA21.1 Protruding ear
LA21.2 Low-set ear
LA21.3 Misshapen ear
LA21.Y Other specified minor anomalies of pinnae
LA22 Structural developmental anomalies of ear causing hearing impairment
LA22.0 Microtia
LA22.1 Anotia
LA22.2 Aplasia or hypoplasia of external auditory canal
LA22.3 Structural developmental anomalies of ear ossicles
LA22.4 Structural developmental anomalies of inner ear
LA22.Y Other specified structural developmental anomalies of ear causing hearing impairment
LA22.Z Structural developmental anomalies of ear causing hearing impairment, unspecified
LA23 Otocephaly

20 Developmental anomalies
LA24 Accessory auricle
LA2Y Other specified structural developmental anomalies of the ear
LA2Z Structural developmental anomalies of the ear, unspecified
Structural developmental anomalies of the face, mouth or teeth
LA30 Structural developmental anomalies of teeth and periodontal tissues
LA30.0 Anodontia
LA30.1 Hypodontia
LA30.2 Oligodontia
LA30.3 Hyperdontia
LA30.4 Abnormalities of size or form of teeth
LA30.5 Anomalies in tooth resorption or loss
LA30.50 Early exfoliation of teeth
LA30.51 Late exfoliation of teeth
DA08.14 Pathological resorption of teeth
LA30.5Y Other specified anomalies in tooth resorption or loss
LA30.5Z Anomalies in tooth resorption or loss, unspecified
LA30.6 Amelogenesis imperfecta
LA30.7 Dentine dysplasia
LA30.8 Dentinogenesis imperfecta
LA30.9 Odontogenesis imperfecta
DA07.3 Disturbances in tooth formation
DA07.4 Root anomaly
DA07.6 Disturbances in tooth eruption
DA07.60 Teething syndrome
DA07.61 Ankylosis of teeth
DA07.6Y Other specified disturbances in tooth eruption
DA07.6Z Disturbances in tooth eruption, unspecified
DA0E.3 Anomalies of tooth position
LA30.Y Other specified structural developmental anomalies of teeth and periodontal tissues
LA30.Z Structural developmental anomalies of teeth and periodontal tissues, unspecified
LA31 Structural developmental anomalies of mouth or tongue
LA31.0 Congenital macroglossia
LA31.1 Hypoglossia or aglossia
LA31.2 Ankyloglossia
LA31.3 Macrostomia
LA31.4 Microstomia
LA31.Y Other specified structural developmental anomalies of mouth or tongue
Clefts of lip, alveolus or palate
LA40 Cleft lip
LA40.0 Cleft lip, unilateral
LA40.1 Cleft lip, bilateral
LA40.2 Cleft lip, median
LA41 Cleft lip and alveolus
LA41.0 Cleft lip and alveolus, unilateral
LA41.1 Cleft lip and alveolus, bilateral
LA42 Cleft palate
LA42.0 Cleft hard palate
LA42.1 Cleft soft palate
LA42.2 Cleft uvula
LA4Y Other specified clefts of lip, alveolus or palate
LA4Z Clefts of lip, alveolus or palate, unspecified

20 Developmental anomalies
LA50 Congenital velopharyngeal incompetence
LA51 Facial clefts
LA52 Facial asymmetry
LA53 Macrocheilia
LA54 Microcheilia
LA55 Compression facies
LA56 Pierre Robin syndrome
LC40 Dermoid cyst
LA5Y Other specified structural developmental anomalies of the face
LA5Z Structural developmental anomalies of the face, unspecified
Structural developmental anomalies of the neck
LA60 Webbed neck
LA61 Congenital sternomastoid tumour
LA62 Congenital torticollis
LA6Y Other specified structural developmental anomalies of the neck
LA6Z Structural developmental anomalies of the neck, unspecified
Structural developmental anomalies of the respiratory system
LA70 Structural developmental anomalies of the nose or cavum
LA70.0 Arrhinia
LA70.1 Bifid nose
LA70.2 Choanal atresia
LA70.3 Congenital perforated nasal septum
LA70.Y Other specified structural developmental anomalies of the nose or cavum
LA70.Z Structural developmental anomalies of the nose or cavum, unspecified
LA71 Structural developmental anomalies of larynx
LA71.0 Congenital laryngomalacia
LA71.1 Laryngocele
LA71.2 Laryngeal hypoplasia
LA71.3 Congenital subglottic stenosis
KB2J Airway obstruction in the neonate due to airway abnormality
KB2J.0 Hypotonia of hypopharynx in neonate
KB2J.1 Hypopharyngeal mass in neonate
KB2J.2 Tracheo-bronchial malacia in neonate
LA31.0 Congenital macroglossia
KB2J.Y Other specified airway obstruction in the neonate due to airway abnormality
KB2J.Z Airway obstruction in the neonate due to airway abnormality, unspecified
LA71.Y Other specified structural developmental anomalies of larynx
LA71.Z Structural developmental anomalies of larynx, unspecified
LA72 Laryngotracheooesophageal cleft
LA73 Structural developmental anomalies of trachea
LA73.0 Congenital stenosis of trachea
LA73.1 Congenital tracheomalacia
LA73.Y Other specified structural developmental anomalies of trachea
LA73.Z Structural developmental anomalies of trachea, unspecified
LA74 Structural developmental anomalies of bronchi
LA74.0 Congenital stenosis or atresia of bronchus
LA74.1 Congenital bronchomalacia
LA74.Y Other specified structural developmental anomalies of bronchi
LA74.Z Structural developmental anomalies of bronchi, unspecified
LA75 Structural developmental anomalies of lungs
LA75.0 Accessory lobe of lung

20 Developmental anomalies

- **LA75.1** Agenesis of lung
- **LA75.2** Congenital hypoplasia of lung
- **LA75.3** Congenital hyperplasia of lung
- **LA75.4** Congenital pulmonary airway malformations
- **LA75.5** Congenital lobar emphysema
- **LA75.6** Congenital sequestration of lung
- **LA75.Y** Other specified structural developmental anomalies of lungs
- **LA75.Z** Structural developmental anomalies of lungs, unspecified

LA76 Structural developmental anomalies of pleura

LA77 Congenital cyst of mediastinum

LA7Y Other specified structural developmental anomalies of the respiratory system

LA7Z Structural developmental anomalies of the respiratory system, unspecified

Structural developmental anomalies of the circulatory system

- Structural developmental anomaly of heart or great vessels
  - **LA80** Anomalous position-orientation of heart
    - **LA80.0** Laevocardia
    - **LA80.1** Dextrocardia
    - **LA80.2** Mesocardia
    - **LA80.3** Extrathoracic heart
    - **LA80.Y** Other specified anomalous position-orientation of heart
    - **LA80.Z** Anomalous position-orientation of heart, unspecified
  - **LA81** Abnormal ventricular relationships
  - **LA82** Total mirror imagery
  - **LA83** Right isomerism
  - **LA84** Left isomerism
  - **LA85** Congenital anomaly of an atrioventricular or ventriculo-arterial connection
    - **LA85.0** Discordant atrioventricular connections
      - **LA85.00** Congenitally corrected transposition of great arteries
      - **LA85.0Y** Other specified discordant atrioventricular connections
      - **LA85.0Z** Discordant atrioventricular connections, unspecified
    - **LA85.1** Transposition of the great arteries
    - **LA85.2** Double outlet right ventricle
      - **LA85.20** Double outlet right ventricle with subpulmonary ventricular septal defect, transposition type
      - **LA85.21** Double outlet right ventricle with non-committed ventricular septal defect
      - **LA85.22** Double outlet right ventricle with subaortic or doubly committed ventricular septal defect without pulmonary stenosis, ventricular septal defect type
      - **LA85.23** Double outlet right ventricle with subaortic or doubly committed ventricular septal defect and pulmonary stenosis, Fallot type
      - **LA85.2Y** Other specified double outlet right ventricle
      - **LA85.2Z** Double outlet right ventricle, unspecified
    - **LA85.3** Double outlet left ventricle
    - **LA85.4** Common arterial trunk
      - **LA85.40** Common arterial trunk with aortic dominance
      - **LA85.41** Common arterial trunk with pulmonary dominance and interrupted aortic arch
      - **LA85.4Y** Other specified common arterial trunk
      - **LA85.4Z** Common arterial trunk, unspecified
    - **LA85.Y** Other specified congenital anomaly of an atrioventricular or ventriculo-arterial connection
    - **LA85.Z** Congenital anomaly of an atrioventricular or ventriculo-arterial connection, unspecified
  - **LA86** Congenital anomaly of mediastinal vein
    - **LA86.0** Left superior caval vein
    - **LA86.1** Unroofed coronary sinus

20 Developmental anomalies
LA86.2 Anomalous pulmonary venous connection
LA86.20 Total anomalous pulmonary venous connection
LA86.21 Partial anomalous pulmonary venous connection
LA86.22 Scimitar syndrome
LA86.2Y Other specified anomalous pulmonary venous connection
LA86.2Z Anomalous pulmonary venous connection, unspecified
LA86.3 Congenital pulmonary venous stenosis or hypoplasia
LA86.Y Other specified congenital anomaly of mediastinal vein
LA86.Z Congenital anomaly of mediastinal vein, unspecified
LA87 Congenital anomaly of an atrioventricular valve or atrioventricular septum
LA87.0 Congenital anomaly of tricuspid valve
LA87.00 Congenital tricuspid regurgitation
LA87.01 Congenital tricuspid valvar stenosis
LA87.02 Dysplasia of tricuspid valve
LA87.03 Ebstein malformation of tricuspid valve
LA87.0Y Other specified congenital anomaly of tricuspid valve
LA87.0Z Congenital anomaly of tricuspid valve, unspecified
LA87.1 Congenital anomaly of mitral valve
LA87.10 Congenital mitral regurgitation
LA87.11 Congenital mitral valvar stenosis
LA87.12 Dysplasia of mitral valve
LA87.13 Congenital anomaly of mitral subvalvar apparatus
LA87.1Y Other specified congenital anomaly of mitral valve
LA87.1Z Congenital anomaly of mitral valve, unspecified
LA87.3 Common atrioventricular junction without an atrioventricular septal defect
LA87.4 Common atrioventricular junction with atrioventricular septal defect
LA87.40 Atrioventricular septal defect with communication at the atrial level only
LA87.41 Atrioventricular septal defect with communication at the ventricular level only
LA87.42 Atrioventricular septal defect with communication at atrial level and restrictive communication at ventricular level
LA87.43 Atrioventricular septal defect with communication at atrial level and unrestrictive communication at ventricular level
LA87.44 Atrioventricular septal defect with ventricular imbalance
LA87.45 Atrioventricular septal defect and tetralogy of Fallot
LA87.4Y Other specified common atrioventricular junction with atrioventricular septal defect
LA87.4Z Common atrioventricular junction with atrioventricular septal defect, unspecified
LA87.Y Other specified congenital anomaly of an atrioventricular valve or atrioventricular septum
LA87.Z Congenital anomaly of an atrioventricular valve or atrioventricular septum, unspecified
LA88 Congenital anomaly of a ventricle or the ventricular septum
LA88.0 Congenital right ventricular outflow tract obstruction
LA88.1 Double chambered right ventricle
LA88.2 Tetralogy of Fallot
LA88.20 Tetralogy of Fallot with absent pulmonary valve syndrome
LA88.21 Tetralogy of Fallot with pulmonary atresia
LA88.22 Tetralogy of Fallot with pulmonary atresia and systemic-to-pulmonary collateral artery
LA88.2Y Other specified tetralogy of Fallot
LA88.2Z Tetralogy of Fallot, unspecified
LA88.3 Congenital left ventricular outflow tract obstruction
LA88.4 Ventricular septal defect
LA88.40 Trabecular muscular ventricular septal defect
LA88.41 Perimembranous central ventricular septal defect

20 Developmental anomalies
LA88.42 Ventricular septal defect haemodynamically insignificant
LA88.4Y Other specified ventricular septal defect
LA88.4Z Ventricular septal defect, unspecified
LA88.Y Other specified congenital anomaly of a ventricle or the ventricular septum
LA88.Z Congenital anomaly of a ventricle or the ventricular septum, unspecified
LA89 Functionally univentricular heart
LA89.0 Double inlet atrioventricular connection
LA89.1 Tricuspid atresia
LA89.2 Mitral atresia
LA89.3 Hypoplastic left heart syndrome
LA89.Y Other specified functionally univentricular heart
LA89.Z Functionally univentricular heart, unspecified
LA8A Congenital anomaly of a ventriculo-arterial valve or adjacent regions
LA8A.0 Congenital anomaly of pulmonary valve
LA8A.00 Congenital pulmonary valvar stenosis
LA8A.01 Congenital pulmonary regurgitation
LA8A.0Y Other specified congenital anomaly of pulmonary valve
LA8A.0Z Congenital anomaly of pulmonary valve, unspecified
LA8A.1 Congenital pulmonary atresia
LA8A.10 Pulmonary atresia with intact ventricular septum
LA8A.1Y Other specified congenital pulmonary atresia
LA8A.1Z Congenital pulmonary atresia, unspecified
LA8A.2 Congenital anomaly of aortic valve
LA8A.20 Congenital aortic valvar stenosis
LA8A.21 Congenital aortic regurgitation
LA8A.22 Bicuspid aortic valve
LA8A.23 Aortic valvar atresia
LA8A.24 Unicuspid aortic valve
LA8A.2Y Other specified congenital anomaly of aortic valve
LA8A.2Z Congenital anomaly of aortic valve, unspecified
LA8A.3 Congenital supravalvar aortic stenosis
LA8A.4 Aneurysm of aortic sinus of Valsalva
LA8A.5 Congenital subaortic stenosis
LA8A.6 Congenital subpulmonary stenosis
LA8A.Y Other specified congenital anomaly of a ventriculo-arterial valve or adjacent regions
LA8A.Z Congenital anomaly of a ventriculo-arterial valve or adjacent regions, unspecified
LA8B Congenital anomaly of great arteries including arterial duct
LA8B.0 Congenital aortopulmonary window
LA8B.1 Congenital anomaly of pulmonary arterial tree
LA8B.2 Congenital anomaly of aorta or its branches
LA8B.21 Coarctation of aorta
LA8B.22 Interrupted aortic arch
LA8B.23 Congenital anomaly of descending thoracic or abdominal aorta
LA8B.24 Congenital anomaly of aortic arch branch
LA8B.2Y Other specified congenital anomaly of aorta or its branches
LA8B.2Z Congenital anomaly of aorta or its branches, unspecified
LA8B.3 Tracheo-oesophageal compressive syndrome
LA8B.4 Patent arterial duct
KB48 Patent arterial duct of prematurity
LA8B.Y Other specified congenital anomaly of great arteries including arterial duct
LA8B.Z Congenital anomaly of great arteries including arterial duct, unspecified

20 Developmental anomalies
LA8C Congenital anomaly of coronary artery
LA8C.0 Anomalous origin of coronary artery from pulmonary arterial tree
LA8C.1 Anomalous aortic origin or course of coronary artery
LA8C.2 Congenital coronary arterial fistula
LA8C.Y Other specified congenital anomaly of coronary artery
LA8C.Z Congenital anomaly of coronary artery, unspecified
LA8D Congenital pericardial anomaly
LA8E Congenital anomaly of atrial septum
LA8E.0 Patent oval foramen
LA8E.1 Atrial septal defect within oval fossa
LA8E.2 Sinus venosus defect
LA8E.3 Interatrial communication through coronary sinus orifice
LA8E.Y Other specified congenital anomaly of atrial septum
LA8E.Z Congenital anomaly of atrial septum, unspecified
LA8F Congenital anomaly of right atrium
LA8G Congenital anomaly of left atrium
LA8G.0 Divided left atrium
LA8G.Y Other specified congenital anomaly of left atrium
LA8G.Z Congenital anomaly of left atrium, unspecified
BD52.6 Congenital great vessel related acquired abnormality
BC02 Acquired abnormality of congenitally malformed valve
BC02.0 Acquired common atrioventricular valvar abnormality in biventricular connections
BC02.1 Acquired truncal valvar abnormality
BC02.2 Acquired common atrioventricular valvar abnormality in double inlet ventricle
BE14.7 Postprocedural common atrioventricular valvar abnormality in double-inlet ventricle
BE14.5 Postprocedural right-sided atrioventricular valvar abnormality in double-inlet ventricle
BE14.6 Postprocedural left-sided atrioventricular valvar abnormality in double-inlet ventricle
BC02.3 Acquired abnormality of neoaortic valve of pulmonary origin
BC02.30 Stenosis of the neoaortic valve of pulmonary origin
BC02.31 Insufficiency of the neoaortic valve of pulmonary origin
BC02.3Y Other specified acquired abnormality of neoaortic valve of pulmonary origin
BC02.3Z Acquired abnormality of neoaortic valve of pulmonary origin, unspecified
BC02.4 Acquired abnormality of the neoaortic valve of truncal origin
BC02.40 Acquired stenosis of the neoaortic valve of truncal origin
BC02.41 Acquired regurgitation of the neoaortic valve of truncal origin
BC02.4Y Other specified acquired abnormality of the neoaortic valve of truncal origin
BC02.4Z Acquired abnormality of the neoaortic valve of truncal origin, unspecified
BC02.Y Other specified acquired abnormality of congenitally malformed valve
BC02.Z Acquired abnormality of congenitally malformed valve, unspecified
BE14.A Postprocedural arterial duct disorder
LA8Y Other specified structural developmental anomaly of heart or great vessels
LA8Z Structural developmental anomaly of heart or great vessels, unspecified
LA90 Structural developmental anomalies of the peripheral vascular system
LA90.0 Capillary malformations
LA90.00 Hereditary haemorrhagic telangiectasia
LC50 Developmental capillary vascular malformations of the skin
LC50.0 Salmon patch
LC50.1 Port-wine stain
LC50.Y Other specified cutaneous capillary vascular malformation
LA90.0Y Other specified capillary malformations
LA90.0Z Capillary malformations, unspecified

20 Developmental anomalies
LA90.1 Lymphatic malformations
LA90.10 Macrocystic lymphatic malformation
LA90.11 Microcystic lymphatic malformation
LA90.12 Lymphatic malformations of certain specified sites
LA90.13 Cystic hygroma in fetus
BD93.0 Primary lymphoedema
EE11.1 Yellow nail syndrome
LD2F.15 Noonan syndrome
LA90.1Y Other specified lymphatic malformations
LA90.1Z Lymphatic malformations, unspecified
LA90.2 Peripheral venous malformations
LA90.20 Vein of Galen aneurysm
LA90.21 Anomalous portal venous connection
LC51 Developmental venous malformations involving the skin
LA90.2Y Other specified peripheral venous malformations
LA90.2Z Peripheral venous malformations, unspecified
LA90.3 Peripheral arteriovenous malformations
LA90.30 Portal vein-hepatic artery fistula
LA90.31 Arteriovenous malformation of precerebral vessels
LA90.32 Uterine arteriovenous malformations
8B22.40 Arteriovenous malformation of cerebral vessels
LA90.3Y Other specified peripheral arteriovenous malformations
LA90.3Z Peripheral arteriovenous malformations, unspecified
LA90.4 Peripheral arterial malformations
LA90.40 Congenital renal artery stenosis
LA90.41 Congenital precerebral nonruptured aneurysm
LA90.42 Congenital cerebral nonruptured aneurysm
8B22.6 Familial cerebral saccular aneurysm
8B22.C Hereditary cerebrovascular diseases
8B22.C0 CADASIL - [cerebral autosomal dominant arteriopathy with subcortical infarcts and leukoencephalopathy] syndrome
8B22.C1 CARASIL - [cerebral autosomal recessive arteriopathy with subcortical infarcts and leukoencephalopathy] syndrome
8B22.CY Other specified hereditary cerebrovascular diseases
8B22.CZ Hereditary cerebrovascular diseases, unspecified
LA90.4Y Other specified peripheral arterial malformations
LA90.4Z Peripheral arterial malformations, unspecified
LA90.5 Pulmonary arteriovenous fistula
LA90.Y Other specified structural developmental anomalies of the peripheral vascular system
LA90.Z Structural developmental anomalies of the peripheral vascular system, unspecified
LA9Y Other specified structural developmental anomalies of the circulatory system
LA9Z Structural developmental anomalies of the circulatory system, unspecified
Structural developmental anomalies of the diaphragm, abdominal wall or umbilical cord
LB00 Structural developmental anomalies of diaphragm
LB00.0 Congenital diaphragmatic hernia
LB00.1 Absence of diaphragm
LB00.Y Other specified structural developmental anomalies of diaphragm
LB00.Z Structural developmental anomalies of diaphragm, unspecified
LB01 Omphalocele
LB02 Gastroschisis
LB03 Structural developmental anomalies of umbilical cord

20 Developmental anomalies
LB03.0 Allantoic duct remnants or cysts
LB03.1 Single umbilical cord artery
2E81.00 Umbilical cord haemangioma
KA03.20 Fetus or newborn affected by short umbilical cord
KA03.21 Fetus or newborn affected by long umbilical cord
EC50 Developmental anomalies of the umbilicus
LB03.0 Allantoic duct remnants or cysts
LB03.Y Other specified structural developmental anomalies of umbilical cord
LB03.Z Structural developmental anomalies of umbilical cord, unspecified
LB0Y Other specified structural developmental anomalies of the diaphragm, abdominal wall or umbilical cord
LB0Z Structural developmental anomalies of the diaphragm, abdominal wall or umbilical cord, unspecified
Structural developmental anomalies of the digestive tract
LB10 Structural developmental anomalies of salivary glands or ducts
LB11 Congenital diverticulum of pharynx
LB12 Structural developmental anomalies of oesophagus
LB12.0 Congenital oesophageal web or ring
LB12.1 Atresia of oesophagus
LB12.10 Atresia of oesophagus with oesophagobronchial fistula
LA72 Laryngotracheooesophageal cleft
LB12.1Y Other specified atresia of oesophagus
LB12.2 Oesophageal fistula without atresia
LB12.3 Congenital stenosis or stricture of oesophagus
LB12.4 Congenital diverticulum of oesophagus
LB12.5 Congenital dilatation of oesophagus
LB12.Y Other specified structural developmental anomalies of oesophagus
LB12.Z Structural developmental anomalies of oesophagus, unspecified
LB13 Structural developmental anomalies of stomach
LB13.0 Congenital hypertrophic pyloric stenosis
LB13.1 Congenital hiatus hernia
LB13.2 Congenital antral web
DA40.2 Gastric volvulus
LB13.Y Other specified structural developmental anomalies of stomach
LB13.Z Structural developmental anomalies of stomach, unspecified
LB14 Structural developmental anomalies of duodenum
LB15 Structural developmental anomalies of small intestine
LB15.0 Meckel diverticulum
LB15.1 Atresia of small intestine
LB15.2 Congenital short bowel
LB15.3 Congenital diverticulitis of small intestine
LB15.4 Congenital diverticulosis of small intestine
LB15.5 Congenital diverticulum of small intestine
LB15.Y Other specified structural developmental anomalies of small intestine
LB15.Z Structural developmental anomalies of small intestine, unspecified
LB16 Structural developmental anomalies of large intestine
LB16.0 Congenital absence, atresia or stenosis of large intestine
LB16.1 Hirschsprung disease
LB16.2 Immature ganglionosis of large intestine
LB16.3 Congenital hypoganglionosis of large intestine
KB87.2 Meconium ileus without perforation
LB16.Y Other specified structural developmental anomalies of large intestine
LB16.Z Structural developmental anomalies of large intestine, unspecified

20 Developmental anomalies
LB17 Structural developmental anomalies of anal canal
LB17.0 Anorectal malformations
LB17.1 Ectopic anus
LB17.2 Persistent cloaca
LB17.3 Cloacal exstrophy
LB17.4 Perineal groove
LB17.Y Other specified structural developmental anomalies of anal canal
LB17.Z Structural developmental anomalies of anal canal, unspecified
LB18 Congenital anomalies of intestinal fixation
DA02.0 Genetic or developmental disorders involving lips or oral mucosa
1F23.14 Chronic mucocutaneous candidosis
LD2D.0 Peutz-Jeghers syndrome
LA90.00 Hereditary haemorrhagic telangiectasia
LB1Y Other specified structural developmental anomalies of the digestive tract
LB1Z Structural developmental anomalies of the digestive tract, unspecified
Structural developmental anomalies of the liver, biliary tract, pancreas or spleen
LB20 Structural developmental anomalies of gallbladder, bile ducts or liver
LB20.0 Structural developmental anomalies of liver
LB20.00 Fibropolycystic liver disease
LB20.20 Choledochal cyst
LB20.21 Biliary atresia
LB20.0Y Other specified structural developmental anomalies of liver
LB20.0Z Structural developmental anomalies of liver, unspecified
LB20.1 Structural developmental anomalies of gallbladder
LB20.10 Agenesis, aplasia or hypoplasia of gallbladder
LB20.1Y Other specified structural developmental anomalies of gallbladder
LB20.1Z Structural developmental anomalies of gallbladder, unspecified
LB20.2 Structural developmental anomalies of bile ducts
LB20.20 Choledochal cyst
LB20.21 Biliary atresia
LB20.22 Congenital stenosis or stricture of bile ducts
LB20.23 Structural developmental anomalies of cystic duct
LB20.24 Accessory bile duct
LB20.2Y Other specified structural developmental anomalies of bile ducts
LB20.2Z Structural developmental anomalies of bile ducts, unspecified
LB20.Y Other specified structural developmental anomalies of gallbladder, bile ducts or liver
LB20.Z Structural developmental anomalies of gallbladder, bile ducts or liver, unspecified
LB21 Structural developmental anomalies of pancreas
LB21.0 Annular pancreas
LB21.1 Pancreas divisum
LB21.2 Accessory pancreas
LB21.3 Agenesis-aplasia of pancreas
LB21.4 Partial agenesis of pancreas
LB21.5 Hypoplasia of pancreas
LB21.Y Other specified structural developmental anomalies of pancreas
LB21.Z Structural developmental anomalies of pancreas, unspecified
LB22 Structural developmental anomalies of spleen
LB22.0 Congenital asplenia
LB22.1 Polysplenia
LB22.2 Ectopic spleen
LB22.Y Other specified structural developmental anomalies of spleen

20 Developmental anomalies
LB22.Z Structural developmental anomalies of spleen, unspecified
LB2Y Other specified structural developmental anomalies of the liver, biliary tract, pancreas or spleen
LB2Z Structural developmental anomalies of the liver, biliary tract, pancreas or spleen, unspecified
Structural developmental anomalies of the urinary system
LB30 Structural developmental anomalies of kidneys
LB30.0 Renal agenesis or other reduction defects of kidney
LB30.00 Renal agenesis
LB30.0Y Other specified renal agenesis or other reduction defects of kidney
LB30.0Z Renal agenesis or other reduction defects of kidney, unspecified
LB30.1 Renal dysplasia
LB30.2 Congenital single renal cyst
LB30.3 Renal tubular dysgenesis
LB30.4 Oligomeganephronia
LB30.5 Accessory kidney
LB30.6 Fusion anomaly of kidneys
LB30.60 Lobulated kidney
LB30.61 Fused pelvic kidney
LB30.62 Horseshoe kidney
LB30.6Y Other specified fusion anomaly of kidneys
LB30.6Z Fusion anomaly of kidneys, unspecified
LB30.7 Ectopic or pelvic kidney
LB30.8 Medullary sponge kidney
LB30.9 Multicystic renal dysplasia
LB30.Y Other specified structural developmental anomalies of kidneys
LB30.Z Structural developmental anomalies of kidneys, unspecified
LB31 Structural developmental anomalies of urinary tract
LB31.0 Congenital hydronephrosis
LB31.1 Congenital primary megaureter
LB31.2 Fetal lower urinary tract obstruction
LB31.3 Exstrophy of urinary bladder
LB31.4 Congenital diverticulum of urinary bladder
LB31.5 Duplication of urethra
LB31.6 Congenital megalourethra
LB31.7 Megacystis-megaureter
LB31.8 Atresia or stenosis of ureter
LB31.9 Agenesis of ureter
LB31.A Duplication of ureter
LB31.B Malposition of ureter
LB31.C Congenital absence of bladder or urethra
LB31.D Congenital vesico-uretero-renal reflux
LB03.0 Allantoic duct remnants or cysts
LB31.Y Other specified structural developmental anomalies of urinary tract
LB31.Z Structural developmental anomalies of urinary tract, unspecified
LB3Y Other specified structural developmental anomalies of the urinary system
LB3Z Structural developmental anomalies of the urinary system, unspecified
Structural developmental anomalies of the female genital system
LB40 Structural developmental anomalies of vulva
LB40.0 Absence of vulva
LB40.1 Embryonic cyst of vulva
LB40.2 Fusion of labia
LB40.Y Other specified structural developmental anomalies of vulva

20 Developmental anomalies
LB40.Z Structural developmental anomalies of vulva, unspecified
LB41 Structural developmental anomalies of clitoris
LB41.0 Agenesis of clitoris
LB41.1 Duplication of clitoris
LB41.2 Clitoromegaly
LB41.Y Other specified structural developmental anomalies of clitoris
LB41.Z Structural developmental anomalies of clitoris, unspecified
LB42 Structural developmental anomalies of vagina
LB42.0 Absence of vagina
LB42.1 Septate vagina
LB42.2 Congenital rectovaginal fistula
LB42.3 Tight hymenal ring
LB42.4 Imperforate hymen
LB42.5 Stricture or atresia of vagina
LB42.Y Other specified structural developmental anomalies of vagina
LB42.Z Structural developmental anomalies of vagina, unspecified
LB43 Structural developmental anomalies of cervix uteri
LB43.0 Embryonic cyst of cervix
LB43.1 Agenesis or aplasia of cervix
LB43.Y Other specified structural developmental anomalies of cervix uteri
LB43.Z Structural developmental anomalies of cervix uteri, unspecified
LB44 Structural developmental anomalies of uterus, except cervix
LB44.0 Agenesis or aplasia of uterine body
LB44.1 Hypoplasia of uterus
LB44.2 Unicornuate uterus
LB44.3 Bicornuate uterus
LB44.4 Septate uterus
LB44.5 Congenital fistulae between uterus and digestive and urinary tracts
LB44.6 Uterovaginal malformation due to diethylstilbestrol syndrome
LB44.Y Other specified structural developmental anomalies of uterus, except cervix
LB44.Z Structural developmental anomalies of uterus, except cervix, unspecified
LB45 Structural developmental anomalies of ovaries, fallopian tubes or broad ligaments
LB45.0 Congenital absence of ovary
LB45.1 46,XX gonadal dysgenesis
LB45.2 Developmental ovarian cyst
LB45.3 Congenital torsion of ovary
LB45.4 Accessory ovary
LB45.5 Congenital absence of fallopian tube
LB45.6 Atresia of fallopian tube
LB45.7 Accessory fallopian tube
LB45.8 Embryonic cyst of fallopian tube
LB45.9 Embryonic cyst of broad ligament
LB45.Y Other specified structural developmental anomalies of ovaries, fallopian tubes or broad ligaments
LB45.Z Structural developmental anomalies of ovaries, fallopian tubes or broad ligaments, unspecified
LB4Y Other specified structural developmental anomalies of the female genital system
LB4Z Structural developmental anomalies of the female genital system, unspecified
Structural developmental anomalies of the male genital system
LB50 Micropenis or penis agenesis
LB51 Anorchia or microorchidia
LD2A.2 Testicular agenesis
LB52 Cryptorchidism

20 Developmental anomalies
LB52.0 Ectopic testis
LB52.1 Undescended testicle, unilateral
LB52.2 Undescended testicle, bilateral
LB52.Y Other specified cryptorchidism
LB52.Z Cryptorchidism, unspecified
LB53 Hypospadias
LB53.0 Hypospadias, balanic
LB53.00 Hypospadias, coronal
LB53.01 Hypospadias, glandular
LB53.0Y Other specified hypospadias, balanic
LB53.0Z Hypospadias, balanic, unspecified
LB53.1 Hypospadias, penile
LB53.2 Hypospadias, penoscrotal
LB53.3 Hypospadias, scrotal
LB53.4 Hypospadias, perineal
LB53.Y Other specified hypospadias
LB53.Z Hypospadias, unspecified
LB54 Congenital chordee
LB55 Epispadias
LB56 Bifid scrotum
LB57 Agenesis of vas deferens
LB58 Polyorchidism
LB59 Hypoplasia of testis or scrotum
LB5Y Other specified structural developmental anomalies of the male genital system
LB5Z Structural developmental anomalies of the male genital system, unspecified
Structural developmental anomalies of the breast
LB60 Breast aplasia
LB61 Absent nipple
LB62 Supernumerary breasts
LB63 Accessory nipple
LB6Y Other specified structural developmental anomalies of the breast
LB6Z Structural developmental anomalies of the breast, unspecified
Structural developmental anomalies of the skeleton
LB70 Structural developmental anomalies of cranium
LB70.0 Craniosynostosis
LB70.00 Plagiocephaly
LB70.0Y Other specified craniosynostosis
LB70.0Z Craniosynostosis, unspecified
LB70.1 Wormian bones
LB70.2 J-shaped sella turcica
LB70.3 Macrocephaly
KD31 Wide cranial sutures of newborn
LB70.Y Other specified structural developmental anomalies of cranium
LB70.Z Structural developmental anomalies of cranium, unspecified
LB71 Structural developmental anomalies of facial bones
LB71.0 Hypotelorism
LB71.1 Hypertelorism
LB71.Y Other specified structural developmental anomalies of facial bones
LB71.Z Structural developmental anomalies of facial bones, unspecified
LB72 Structural developmental anomalies of shoulder girdle
LB72.0 Cervical rib

20 Developmental anomalies
LB72.1 Sprengel deformity
LB72.2 Deformation of scapula
LB72.Y Other specified structural developmental anomalies of shoulder girdle
LB72.Z Structural developmental anomalies of shoulder girdle, unspecified
LB73 Structural developmental anomalies of spine or bony thorax
LB73.0 Occult spinal dysraphism
LB73.1 Structural developmental anomalies of chest wall
LB73.10 Poland syndrome
LB73.11 Bifid rib
LB73.12 Accessory rib
LB73.13 Structural developmental anomalies of sternum
LB73.1Y Other specified structural developmental anomalies of chest wall
LB73.1Z Structural developmental anomalies of chest wall, unspecified
LB73.2 Structural developmental anomalies of spine
LB73.20 Klippel-Feil anomaly
LB73.21 Occipitalisation of atlas
LB73.22 Atlanto-axial instability or subluxation
LB73.23 Aplasia or hypoplasia of the odontoid process of axis
LB73.24 Segmentation anomalies of vertebrae
LB73.25 Congenital scoliosis due to congenital bony malformation
LB73.26 Sacralization of the last lumbar vertebra
LB73.27 Lumbarisation of the first sacral vertebra
LB73.28 Sacrum agenesis or hypoplasia
LB73.29 Caudal appendage
LB73.2A Congenital spondylolisthesis
LB73.2Y Other specified structural developmental anomalies of spine
LB73.2Z Structural developmental anomalies of spine, unspecified
LB73.Y Other specified structural developmental anomalies of spine or bony thorax
LB73.Z Structural developmental anomalies of spine or bony thorax, unspecified
LB74 Structural developmental anomalies of pelvic girdle
LB74.0 Developmental dysplasia of hip
LB74.1 Congenital subluxation of hip
LB74.2 Unstable hip
LB74.3 Congenital coxa vara
LB74.4 Congenital coxa valga
LB74.5 Wide symphysis pubis
LB74.Y Other specified structural developmental anomalies of pelvic girdle
LB74.Z Structural developmental anomalies of pelvic girdle, unspecified
LB75 Brachydactyly
LB75.0 Brachydactyly of fingers
LB75.1 Brachydactyly of toes
LB75.2 Symbrachydactyly of hands or feet
LB75.Y Other specified brachydactyly
LB75.Z Brachydactyly, unspecified
LB76 Triphalangeal thumb
LB77 Hyperphalangy
LB78 Polydactyly
LB78.0 Polydactyly of the thumb
LB78.1 Polysyndactyly
LB78.2 Postaxial polydactyly of fingers
LB78.3 Polydactyly of toes

20 Developmental anomalies
LB78.Y Other specified polydactyly
LB78.Z Polydactyly, unspecified
LB79 Syndactyly
LB79.0 Fused fingers
LB79.1 Webbed fingers
LB79.2 Fused toes
LB79.3 Webbed toes
LB78.1 Polysyndactyly
LB79.Y Other specified syndactyly
LB79.Z Syndactyly, unspecified
Congenital deformities of fingers or toes
LB80 Congenital deformities of fingers
LB80.0 Clinodactyly of fingers
LB80.2 Radial deviation of fingers
LB80.Y Other specified congenital deformities of fingers
LB81 Congenital deformities of toes
LB81.0 Clinodactyly of toes
LB98.5 Congenital hammer toe
LB81.Y Other specified congenital deformities of toes
LB8Z Congenital deformities of fingers or toes, unspecified
LB90 Joint formation defects
LB90.0 Humero-radio-ulnar synostosis
LB90.1 Humero-radial synostosis
LB90.2 Humero-ulnar synostosis
LB90.3 Radio-ulnar synostosis
LB90.4 Madelung deformity
LB90.5 Congenital digital clubbing
LB90.6 Tibio-fibular synostosis
LB90.7 Cubitus valgus
LB90.8 Cubitus varus
LB90.Y Other specified joint formation defects
LB90.Z Joint formation defects, unspecified
LB91 Congenital shoulder dislocation
LB92 Congenital elbow dislocation
LB93 Congenital knee dislocation
LB93.0 Congenital genu recurvatum
LB93.1 Congenital genu flexum
LB93.Y Other specified congenital knee dislocation
LB93.Z Congenital knee dislocation, unspecified
LB94 Congenital patella dislocation
LB95 Patella aplasia or hypoplasia
LB96 Congenital bowing of long bones
LB96.0 Congenital bowing of femur
LB96.1 Congenital bowing of tibia
LB96.Y Other specified congenital bowing of long bones
LB96.Z Congenital bowing of long bones, unspecified
LB97 Limb overgrowth
LB97.0 Macrodactyly of fingers
LB97.1 Macrodactyly of toes
LB97.2 Upper limb hypertrophy
LB97.3 Lower limb hypertrophy

20 Developmental anomalies
LB97.Y Other specified limb overgrowth
LB97.Z Limb overgrowth, unspecified
LB98 Congenital deformities of feet
LB98.0 Congenital varus deformities of feet
LB98.00 Talipes equinovarus
LB98.01 Talipes calcaneovarus
LB98.02 Metatarsus varus
LB98.0Y Other specified congenital varus deformities of feet
LB98.0Z Congenital varus deformities of feet, unspecified
LB98.1 Congenital pes planus
LB98.2 Congenital valgus deformities of feet
LB98.20 Congenital hallux valgus
LB98.21 Metatarsus valgus
LB98.22 Talipes calcaneovalgus
LB98.2Y Other specified congenital valgus deformities of feet
LB98.2Z Congenital valgus deformities of feet, unspecified
LB98.3 Congenital pes cavus
LB98.4 Congenital vertical talus
LB98.5 Congenital hammer toe
LB98.Y Other specified congenital deformities of feet
LB98.Z Congenital deformities of feet, unspecified
LB99 Reduction defects of upper limb
LB99.0 Amelia of upper limb
LB99.1 Humeral agenesis or hypoplasia
LB99.2 Radial hemimelia
LB99.3 Ulnar hemimelia
LB99.4 Congenital absence of upper arm or forearm with hand present
LB99.5 Congenital absence of both forearm and hand
LB99.6 Acheiria
LB99.7 Adactyly of hands
LB99.8 Split hand
LB99.Y Other specified reduction defects of upper limb
LB99.Z Reduction defects of upper limb, unspecified
LB9A Reduction defects of lower limb
LB9A.0 Amelia of lower limb
LB9A.1 Tibial hemimelia
LB9A.2 Fibular hemimelia
LB9A.3 Congenital absence of thigh or lower leg with foot present
LB9A.4 Apodia
LB9A.5 Adactyly of feet
LB9A.6 Split foot
LB9A.7 Congenital absence of both lower leg and foot
LB9A.8 Femoral agenesis or hypoplasia
LB9A.Y Other specified reduction defects of lower limb
LB9A.Z Reduction defects of lower limb, unspecified
LB9B Reduction defects of upper and lower limbs
LB9Y Other specified structural developmental anomalies of the skeleton
LB9Z Structural developmental anomalies of the skeleton, unspecified
Structural developmental anomalies of the skin
Developmental hamartomata of the epidermis and epidermal appendages
LC00 Keratinocytic epidermal hamartoma

20 Developmental anomalies
LC00.0 Epidermal naevus
LC00.Y Other specified keratinocytic epidermal hamartoma
LC01 Pilosebaceous hamartoma
LC02 Complex epidermal hamartoma
LC0Y Other specified developmental hamartomata of the epidermis and epidermal appendages
Developmental anomalies of skin pigmentation
LC10 Dermal melanocytosis
2F20.2 Congenital melanocytic naevus
2F20.20 Giant congenital melanocytic naevus
2F20.2Y Other specified congenital melanocytic naevus
2F20.2Z Congenital melanocytic naevus, unspecified
LC1Y Other specified developmental anomalies of skin pigmentation
Hamartomata derived from dermal connective tissue
LC20 Connective tissue hamartoma
LC2Y Other specified hamartomata derived from dermal connective tissue
Developmental defects of hair or nails
LC30 Developmental defects of hair or hair growth
LC31 Developmental defects of the nail apparatus
LC40 Dermoid cyst
Developmental anomalies of cutaneous vasculature
LC50 Developmental capillary vascular malformations of the skin
LC51 Developmental venous malformations involving the skin
LC52 Complex or combined developmental vascular malformations involving the skin
LD26.60 Angio-osteohypertrophic syndrome
LC5Y Other specified developmental anomalies of cutaneous vasculature
LC5Z Developmental anomalies of cutaneous vasculature, unspecified
Congenital anomalies of skin development
LC60 Aplasia cutis congenita
LC7Y Other specified structural developmental anomalies of the skin
LC7Z Structural developmental anomalies of the skin, unspecified
Structural developmental anomalies of the adrenal glands
LC80 Congenital adrenal hypoplasia
LC8Y Other specified structural developmental anomalies of the adrenal glands
LC8Z Structural developmental anomalies of the adrenal glands, unspecified
LD0Y Other specified structural developmental anomalies primarily affecting one body system
LD0Z Structural developmental anomalies primarily affecting one body system, unspecified
Multiple developmental anomalies or syndromes
LD20 Syndromes with central nervous system anomalies as a major feature
LD20.0 Syndromes with cerebellar anomalies as a major feature
LD20.00 Joubert syndrome
LD20.01 Pontocerebellar hypoplasia
LD20.0Y Other specified syndromes with cerebellar anomalies as a major feature
LD20.0Z Syndromes with cerebellar anomalies as a major feature, unspecified
LD20.1 Syndromes with lissencephaly as a major feature
LD20.2 Syndromes with microcephaly as a major feature
LD20.3 Syndromes with holoprosencephaly as a major feature
LA05.4 Arrhinencephaly
LD20.4 Syndromes with brain calcifications as a major feature
LD20.Y Other specified syndromes with central nervous system anomalies as a major feature
LD20.Z Syndromes with central nervous system anomalies as a major feature, unspecified
LD21 Syndromes with eye anomalies as a major feature

20 Developmental anomalies
LD21.0 Syndromes with microphthalmia as a major feature
LD21.Y Other specified syndromes with eye anomalies as a major feature
LD21.Z Syndromes with eye anomalies as a major feature, unspecified
LD22 Syndromes with dental anomalies as a major feature
LA30.6 Amelogenesis imperfecta
LA30.7 Dentine dysplasia
LA30.8 Dentinogenesis imperfecta
LD23 Syndromes with vascular anomalies as a major feature
LD26.60 Angio-osteohypertrophic syndrome
BD93.0 Primary lymphoedema
LD2F.15 Noonan syndrome
LD24 Syndromes with skeletal anomalies as a major feature
LD24.0 Syndromes with micromelia
LD24.00 Achondroplasia
LD24.01 Hypochondroplasia
LD24.02 Thanatophoric dysplasia
LD24.03 Diastrophic dysplasia
LD24.04 Chondrodysplasia punctata
LD24.0Y Other specified syndromes with micromelia
LD24.0Z Syndromes with micromelia, unspecified
LD24.1 Bone diseases with increased bone density
LD24.10 Osteopetrosis
LD24.11 Osteopoikilosis
LD24.1Y Other specified bone diseases with increased bone density
LD24.1Z Bone diseases with increased bone density, unspecified
LD24.2 Bone diseases with disorganised development of skeletal components
LD24.20 Multiple osteochondromas
LD24.21 Exostoses with anetodermia and brachydactyly type E
LD24.22 Cherubism
LD24.23 Yunis-Varon disease
LD24.K0 Osteogenesis imperfecta
LD24.2Y Other specified bone diseases with disorganised development of skeletal components
LD24.2Z Bone diseases with disorganised development of skeletal components, unspecified
LD24.3 Spondyloepiphyseal or spondyloepimetaphyseal dysplasias
LD24.4 Spondylometaphyseal dysplasias
LD24.5 Spondylodysplastic dysplasias
LD24.50 Achondrogenesis
LD24.51 Hypochondrogenesis
LD24.5Y Other specified spondylodysplastic dysplasias
LD24.5Z Spondylodysplastic dysplasias, unspecified
LD24.6 Multiple epiphyseal dysplasia or pseudoachondroplasia
LD24.60 Pseudoachondroplasia
LD24.61 Multiple epiphyseal dysplasias
LD24.6Y Other specified multiple epiphyseal dysplasia or pseudoachondroplasia
LD24.6Z Multiple epiphyseal dysplasia or pseudoachondroplasia, unspecified
LD24.7 Multiple metaphyseal dysplasias
LD24.8 Acromelic dysplasias
LD24.80 Langer-Giedion syndrome
LD24.8Y Other specified acromelic dysplasias
LD24.8Z Acromelic dysplasias, unspecified
LD24.9 Acromesomelic dysplasias

20 Developmental anomalies
LD24.A Mesomelic or rhizomesomelic dysplasias
LD24.B Short rib syndromes
LD24.B0 Short rib-polydactyly syndrome
LD24.B1 Asphyxiating thoracic dystrophy
LD24.BY Other specified short rib syndromes
LD24.BZ Short rib syndromes, unspecified
LD24.C Bent bone dysplasias
LD24.D Slender bone dysplasias
LD24.E Bone dysplasias with multiple joint dislocations
LD24.F Progressive ossification of skin, skeletal muscle, fascia, tendons or ligaments
FB31.0 Progressive osseous heteroplasia
FB31.1 Fibrodysplasia ossificans progressiva
LD24.G Syndromic craniosynostoses
LD24.G0 Pfeiffer syndrome
LD24.G1 Crouzon disease
LD24.G2 Apert syndrome
LD24.GY Other specified syndromic craniosynostoses
LD24.GZ Syndromic craniosynostoses, unspecified
LD24.H Dysostoses with predominant vertebral and costal involvement
LD24.J Patellar dysostoses
LD24.J0 Nail-patella syndrome
LD24.JY Other specified patellar dysostoses
LD24.JZ Patellar dysostoses, unspecified
LD24.K Genetic bone diseases with decreased bone density
LD24.K0 Osteogenesis imperfecta
LD24.KY Other specified genetic bone diseases with decreased bone density
LD24.KZ Genetic bone diseases with decreased bone density, unspecified
FB31.0 Progressive osseous heteroplasia
FB31.1 Fibrodysplasia ossificans progressiva
LD24.Y Other specified syndromes with skeletal anomalies as a major feature
LD24.Z Syndromes with skeletal anomalies as a major feature, unspecified
LD25 Syndromes with face or limb anomalies as a major feature
LD25.0 Oromandibular-limb anomaly syndrome
LD25.00 Oral-facial-digital syndrome
LD25.0Y Other specified oromandibular-limb anomaly syndrome
LD25.0Z Oromandibular-limb anomaly syndrome, unspecified
LD25.1 Fronto-otopalatodigital syndromes
LD25.2 Acrofacial dysostoses
LD25.3 Craniofacial dysostoses
LD25.Y Other specified syndromes with face or limb anomalies as a major feature
LD25.Z Syndromes with face or limb anomalies as a major feature, unspecified
LD26 Syndromes with limb anomalies as a major feature
LD26.0 Combined reduction defects of upper and lower limbs
LD26.1 Complex brachydactylies
LD26.2 Syndromes with limb duplication, polydactyly, syndactyly or triphalangism
LD26.3 Syndromes with synostoses of limbs
LD26.4 Arthrogryposis syndromes
LD26.40 Multiple pterygium syndrome
LD26.41 Arthrogryposis multiplex congenita
LD26.4Y Other specified arthrogryposis syndromes
LD26.4Z Arthrogryposis syndromes, unspecified
LD26.5 Constriction rings

20 Developmental anomalies
LD26.6 Congenital vascular bone syndromes
LD26.60 Angio-osteohypertrophic syndrome
LD26.6Y Other specified congenital vascular bone syndromes
LD26.6Z Congenital vascular bone syndromes, unspecified
LD26.Y Other specified syndromes with limb anomalies as a major feature
LD26.Z Syndromes with limb anomalies as a major feature, unspecified
LD27 Syndromes with skin or mucosal anomalies as a major feature
LD27.0 Ectodermal dysplasia syndromes
LD27.00 Incontinentia pigmenti
LD27.01 Cronkhite-Canada syndrome
LD27.02 Hypohidrotic ectodermal dysplasia
LD27.03 Hidrotic ectodermal dysplasia, Clouston type
LD24.80 Langer-Giedion syndrome
LD25.00 Oral-facial-digital syndrome
LD27.0Y Other specified ectodermal dysplasia syndromes
LD27.1 Xeroderma pigmentosum
LD27.2 Syndromic ichthyosis
5C52.03 Sjögren-Larsson syndrome
LD27.3 Genetic syndromes with hypertrichosis
LD27.4 Genetic syndromes affecting nails
5C64.00 Wilson disease
LD24.J0 Nail-patella syndrome
LD27.03 Hidrotic ectodermal dysplasia, Clouston type
LD27.5 Genetic hamartoneoplastic syndromes affecting the skin
LD2D.1 Neurofibromatoses
LD2D.2 Tuberous sclerosis
LD2D.3 Gardner syndrome
LD2D.4 Gorlin syndrome
LD27.6 Genetic lipodystrophy
LD27.60 Congenital generalised lipodystrophy
LD27.6Z Genetic lipodystrophy, unspecified
5C64.20 Acrodermatitis enteropathica
EC20.0 Non-syndromic ichthyosis
EC20.00 Ichthyosis vulgaris
EC20.01 X-linked ichthyosis
EC20.02 Autosomal recessive congenital ichthyosis
EC20.03 Keratinopathic ichthyoses
EC20.0Y Other specified or unclassifiable non-syndromic ichthyosis
EC40 Pseudoxanthoma elasticum
LD27.Y Other specified syndromes with skin or mucosal anomalies as a major feature
LD27.Z Syndromes with skin or mucosal anomalies as a major feature, unspecified
LD28 Syndromes with connective tissue involvement as a major feature
LD28.0 Marfan syndrome or Marfan-related disorders
LD28.00 Congenital contractural arachnodactyly
LD28.01 Marfan syndrome
LD28.0Y Other specified Marfan syndrome or Marfan-related disorders
LD28.0Z Marfan syndrome or Marfan-related disorders, unspecified
LD28.1 Ehlers-Danlos syndrome
LD28.10 Ehlers-Danlos syndrome, classical type
LD28.1Y Other specified types of Ehlers-Danlos syndrome
LD28.2 Genetically-determined cutis laxa
LD28.Y Other specified syndromes with connective tissue involvement as a major feature

20 Developmental anomalies
LD28.Z Syndromes with connective tissue involvement as a major feature, unspecified
LD29 Syndromes with obesity as a major feature
LD90.3 Prader-Willi syndrome
LD2A Malformative disorders of sex development
LD2A.0 Ovotesticular disorder of sex development
LD2A.1 46,XY gonadal dysgenesis
LD2A.2 Testicular agenesis
LD2A.3 46,XY disorder of sex development due to a defect in testosterone metabolism
LD2A.4 46,XY disorder of sex development due to androgen resistance
LD56 Chimaera 46, XX, 46, XY
LD56.0 Androgenetic chimaera
LD56.1 Gynogenetic chimaera
LD56.Y Other specified chimaera 46, XX, 46, XY
LD56.Z Chimaera 46, XX, 46, XY, unspecified
5A71.1 46,XX disorders of sex development induced by androgens of maternal origin
5A71.01 Congenital adrenal hyperplasia
LD2A.Y Other specified malformative disorders of sex development
LD2A.Z Malformative disorders of sex development, unspecified
LD2B Syndromes with premature ageing appearance as a major feature
LD2C Overgrowth syndromes
LD2D Phakomatoses or hamartoneoplastic syndromes
LD2D.0 Peutz-Jeghers syndrome
LD2D.1 Neurofibromatoses
LD2D.2 Tuberous sclerosis
LD2D.3 Gardner syndrome
LD2D.4 Gorlin syndrome
LD2D.Y Other specified phakomatoses or hamartoneoplastic syndromes
LD2D.Z Phakomatoses or hamartoneoplastic syndromes, unspecified
LD2E Syndromes with structural anomalies due to inborn errors of metabolism
5C52.10 Disorders of cholesterol synthesis
5C53.02 Pyruvate dehydrogenase complex deficiency
5C54 Inborn errors of glycosylation or other specified protein modification
5C54.0 Disorders of protein N-glycosylation
5C54.1 Disorders of protein O-glycosylation
LD24.20 Multiple osteochondromas
5C54.2 Disorders of multiple glycosylation or other pathways
5C54.Y Other specified congenital disorders of glycosylation and protein modification
5C54.Z Congenital disorders of glycosylation and protein modification, unspecified
5C56.01 Fabry disease
5C56.20 Mucolipidosis
5C56.21 Oligosaccharidosis
5C56.3 Mucopolysaccharidosis
5C56.30 Mucopolysaccharidosis type 1
5C56.31 Mucopolysaccharidosis type 2
5C56.32 Mucopolysaccharidosis type 4
5C56.33 Mucopolysaccharidosis type 6
5C56.3Y Other specified mucopolysaccharidosis
5C56.3Z Mucopolysaccharidosis, unspecified
LD2F Syndromes with multiple structural anomalies, without predominant body system involvement
LD2F.0 Toxic or drug-related embryofetopathies
LD2F.00 Fetal alcohol syndrome
LD2F.01 Fetal hydantoin syndrome

20 Developmental anomalies
LD2F.02 Embryofetopathy due to oral anticoagulant therapy
LD2F.03 Fetal Valproate Spectrum Disorder
LB44.6 Uterovaginal malformation due to diethylstilbestrol syndrome
LD2F.0Y Other specified toxic or drug-related embryofetopathies
LD2F.0Z Toxic or drug-related embryofetopathies, unspecified
LD2F.1 Syndromes with multiple structural anomalies, not of environmental origin
LD2F.10 Prune belly syndrome
LD2F.11 VATER association
LD2F.12 Sirenomelia
LD2F.13 Meckel-Gruber syndrome
LD2F.14 MURCS association
LD2F.15 Noonan syndrome
LD2F.16 Otomandibular dysplasia
LD2H.0 Fraser syndrome
LD2H.3 Waardenburg-Shah syndrome
5C60.0 Oculocerebrorenal syndrome
LD2F.1Y Other specified syndromes with multiple structural anomalies, not of environmental origin
LD2F.1Z Syndromes with multiple structural anomalies, not of environmental origin, unspecified
KA62.8 Congenital rubella syndrome
KA62.3 Congenital cytomegalovirus infection
KA62.A Perinatal Herpes simplex infection
KA62.1 Congenital Epstein-Barr virus infection
KA62.7 Congenital parvovirus syndrome
KA62.5 Congenital enterovirus infection
KA64.0 Congenital toxoplasmosis
KA62.0 Congenital Zika virus infection
KA62.2 Congenital Varicella Zoster virus infection
5C50.02 Embryofetopathy due to maternal phenylketonuria
LD2F.Y Other specified syndromes with multiple structural anomalies, without predominant body system involvement
LD2F.Z Syndromes with multiple structural anomalies, without predominant body system involvement, unspecified
LD2G Conjoined twins
LD2H Syndromic genetic deafness
LD2H.0 Fraser syndrome
LD2H.1 Neuropathy with hearing impairment
LD2H.2 Progressive deafness with stapes fixation
LD2H.3 Waardenburg-Shah syndrome
LD2H.4 Usher syndrome
LD44.N0 CATCH 22 phenotype
5A00.02 Pendred syndrome
5A05 Generalised resistance to thyroid hormone
LD2H.Y Other specified syndromic genetic deafness
LD2H.Z Syndromic genetic deafness, unspecified
LD2Y Other specified multiple developmental anomalies or syndromes
LD2Z Multiple developmental anomalies or syndromes, unspecified
Chromosomal anomalies, excluding gene mutations
LD40 Complete trisomies of the autosomes
LD40.0 Complete trisomy 21
LD40.1 Complete trisomy 13
LD40.2 Complete trisomy 18
LD40.Y Other specified complete trisomies of the autosomes

20 Developmental anomalies
LD40.Z Complete trisomies of the autosomes, unspecified
LD41 Duplications of the autosomes
LD41.0 Duplications of chromosome 1
LD41.00 Duplications of the long arm of chromosome 1
LD41.01 Duplications of the short arm of chromosome 1
LD41.0Y Other specified duplications of chromosome 1
LD41.0Z Duplications of chromosome 1, unspecified
LD41.1 Duplications of chromosome 2
LD41.10 Duplications of the long arm of chromosome 2
LD41.11 Duplications of the short arm of chromosome 2
LD41.1Y Other specified duplications of chromosome 2
LD41.1Z Duplications of chromosome 2, unspecified
LD41.2 Duplications of chromosome 3
LD41.20 Duplications of the long arm of chromosome 3
LD41.21 Duplications of the short arm of chromosome 3
LD41.2Y Other specified duplications of chromosome 3
LD41.2Z Duplications of chromosome 3, unspecified
LD41.3 Duplications of chromosome 4
LD41.30 Duplications of the long arm of chromosome 4
LD41.31 Duplications of the short arm of chromosome 4
LD41.3Y Other specified duplications of chromosome 4
LD41.3Z Duplications of chromosome 4, unspecified
LD41.4 Duplications of chromosome 5
LD41.40 Duplications of the long arm of chromosome 5
LD41.41 Duplications of the short arm of chromosome 5
LD41.4Y Other specified duplications of chromosome 5
LD41.4Z Duplications of chromosome 5, unspecified
LD41.5 Duplications of chromosome 6
LD41.50 Duplications of the long arm of chromosome 6
LD41.51 Duplications of the short arm of chromosome 6
LD41.5Y Other specified duplications of chromosome 6
LD41.5Z Duplications of chromosome 6, unspecified
LD41.6 Duplications of chromosome 7
LD41.60 Duplications of the long arm of chromosome 7
LD41.61 Duplications of the short arm of chromosome 7
LD41.6Y Other specified duplications of chromosome 7
LD41.6Z Duplications of chromosome 7, unspecified
LD41.7 Duplications of chromosome 8
LD41.70 Duplications of the long arm of chromosome 8
LD41.71 Duplications of the short arm of chromosome 8
LD41.7Y Other specified duplications of chromosome 8
LD41.7Z Duplications of chromosome 8, unspecified
LD41.8 Duplications of chromosome 9
LD41.80 Duplications of the long arm of chromosome 9
LD41.81 Duplications of the short arm of chromosome 9
LD41.8Y Other specified duplications of chromosome 9
LD41.8Z Duplications of chromosome 9, unspecified
LD41.9 Duplications of chromosome 10
LD41.90 Duplications of the long arm of chromosome 10
LD41.91 Duplications of the short arm of chromosome 10
LD41.9Y Other specified duplications of chromosome 10
LD41.9Z Duplications of chromosome 10, unspecified

20 Developmental anomalies
LD41.A Duplications of chromosome 11
LD41.B Duplications of chromosome 12
LD41.B0 Duplications of the long arm of chromosome 12
LD41.B1 Duplications of the short arm of chromosome 12
LD41.BY Other specified duplications of chromosome 12
LD41.BZ Duplications of chromosome 12, unspecified
LD41.C Duplications of chromosome 13
LD41.D Duplications of chromosome 14
LD41.E Duplications of chromosome 15
LD41.F Duplications of chromosome 16
LD41.F0 Duplications of the long arm of chromosome 16
LD41.F1 Duplications of the short arm of chromosome 16
LD41.FY Other specified duplications of chromosome 16
LD41.FZ Duplications of chromosome 16, unspecified
LD41.G Duplications of chromosome 17
LD41.G0 Duplications of the long arm of chromosome 17
LD41.G1 Duplications of the short arm of chromosome 17
LD41.GY Other specified duplications of chromosome 17
LD41.GZ Duplications of chromosome 17, unspecified
LD41.H Duplications of chromosome 18
LD41.H0 Duplications of the long arm of chromosome 18
LD41.H1 Duplications of the short arm of chromosome 18
LD41.HY Other specified duplications of chromosome 18
LD41.HZ Duplications of chromosome 18, unspecified
LD41.J Duplications of chromosome 19
LD41.J0 Duplications of the long arm of chromosome 19
LD41.J1 Duplications of the short arm of chromosome 19
LD41.JY Other specified duplications of chromosome 19
LD41.JZ Duplications of chromosome 19, unspecified
LD41.K Duplications of chromosome 20
LD41.K0 Duplications of the long arm of chromosome 20
LD41.K1 Duplications of the short arm of chromosome 20
LD41.KY Other specified duplications of chromosome 20
LD41.KZ Duplications of chromosome 20, unspecified
LD41.L Duplications of chromosome 21
LD41.M Duplications of chromosome 22
LD41.N Extra ring or dicentric chromosomes
LD41.P Duplications with other complex rearrangements
LD41.Q Extra marker chromosomes
LD41.Y Other specified duplications of the autosomes
LD41.Z Duplications of the autosomes, unspecified
LD42 Polyploidies
LD42.0 Triploidy
LD42.1 Tetraploidy
LD42.Y Other specified polyploidies
LD42.Z Polyploidies, unspecified
LD43 Complete monosomies of the autosomes
LD43.0 Complete monosomy of autosome
LD43.1 Mosaic monosomy of autosome
LD43.Y Other specified complete monosomies of the autosomes
LD43.Z Complete monosomies of the autosomes, unspecified
LD44 Deletions of the autosomes

20 Developmental anomalies
LD44.1 Deletions of chromosome 1
LD44.10 Deletions of the long arm of chromosome 1
LD44.11 Deletions of the short arm of chromosome 1
LD44.1Y Other specified deletions of chromosome 1
LD44.1Z Deletions of chromosome 1, unspecified
LD44.2 Deletions of chromosome 2
LD44.20 Deletions of the long arm of chromosome 2
LD44.21 Deletions of the short arm of chromosome 2
LD44.2Y Other specified deletions of chromosome 2
LD44.2Z Deletions of chromosome 2, unspecified
LD44.3 Deletions of chromosome 3
LD44.30 Deletions of the long arm of chromosome 3
LD44.31 Deletions of the short arm of chromosome 3
LD44.3Y Other specified deletions of chromosome 3
LD44.3Z Deletions of chromosome 3, unspecified
LD44.4 Deletions of chromosome 4
LD44.40 Deletions of the long arm of chromosome 4
LD44.41 Deletions of the short arm of chromosome 4
LD44.4Y Other specified deletions of chromosome 4
LD44.4Z Deletions of chromosome 4, unspecified
LD44.5 Deletions of chromosome 5
LD44.50 Deletions of the long arm of chromosome 5
LD44.51 Deletions of the short arm of chromosome 5
LD44.5Y Other specified deletions of chromosome 5
LD44.5Z Deletions of chromosome 5, unspecified
LD44.6 Deletions of chromosome 6
LD44.60 Deletions of the long arm of chromosome 6
LD44.61 Deletions of the short arm of chromosome 6
LD44.6Y Other specified deletions of chromosome 6
LD44.6Z Deletions of chromosome 6, unspecified
LD44.7 Deletions of chromosome 7
LD44.70 Deletions of the long arm of chromosome 7
LD44.71 Deletions of the short arm of chromosome 7
LD44.7Y Other specified deletions of chromosome 7
LD44.7Z Deletions of chromosome 7, unspecified
LD44.8 Deletions of chromosome 8
LD44.80 Deletions of the long arm of chromosome 8
LD24.80 Langer-Giedion syndrome
LD44.81 Deletions of the short arm of chromosome 8
LD44.8Y Other specified deletions of chromosome 8
LD44.8Z Deletions of chromosome 8, unspecified
LD44.9 Deletions of chromosome 9
LD44.90 Deletions of the long arm of chromosome 9
LD44.91 Deletions of the short arm of chromosome 9
LD44.9Y Other specified deletions of chromosome 9
LD44.9Z Deletions of chromosome 9, unspecified
LD44.A Deletions of chromosome 10
LD44.A0 Deletions of the long arm of chromosome 10
LD44.A1 Deletions of the short arm of chromosome 10
LD44.AY Other specified deletions of chromosome 10
LD44.AZ Deletions of chromosome 10, unspecified
LD44.B Deletions of chromosome 11

20 Developmental anomalies
LD44.B0 Deletions of the long arm of chromosome 11
LD44.B1 Deletions of the short arm of chromosome 11
LD44.BY Other specified deletions of chromosome 11
LD44.BZ Deletions of chromosome 11, unspecified
LD44.C Deletions of chromosome 12
LD44.C0 Deletions of the long arm of chromosome 12
LD44.C1 Deletions of the short arm of chromosome 12
LD44.CY Other specified deletions of chromosome 12
LD44.CZ Deletions of chromosome 12, unspecified
LD44.D Deletions of chromosome 13
LD44.E Deletions of chromosome 14
LD44.F Deletions of chromosome 15
LD44.G Deletions of chromosome 16
LD44.G0 Deletions of the long arm of chromosome 16
LD44.G1 Deletions of the short arm of chromosome 16
LD44.GY Other specified deletions of chromosome 16
LD44.GZ Deletions of chromosome 16, unspecified
LD44.H Deletions of chromosome 17
LD44.H0 Deletions of the long arm of chromosome 17
LD44.H1 Deletions of the short arm of chromosome 17
LD44.HY Other specified deletions of chromosome 17
LD44.HZ Deletions of chromosome 17, unspecified
LD44.J Deletions of chromosome 18
LD44.J0 Deletions of the long arm of chromosome 18
LD44.J1 Deletions of the short arm of chromosome 18
LD44.JY Other specified deletions of chromosome 18
LD44.JZ Deletions of chromosome 18, unspecified
LD44.K Deletions of chromosome 19
LD44.K0 Deletions of the long arm of chromosome 19
LD44.K1 Deletions of the short arm of chromosome 19
LD44.KY Other specified deletions of chromosome 19
LD44.KZ Deletions of chromosome 19, unspecified
LD44.L Deletions of chromosome 20
LD44.L0 Deletions of the long arm of chromosome 20
LD44.L1 Deletions of the short arm of chromosome 20
LD44.LY Other specified deletions of chromosome 20
LD44.LZ Deletions of chromosome 20, unspecified
LD44.M Deletions of chromosome 21
LD44.N Deletions of chromosome 22
LD44.N0 CATCH 22 phenotype
LD44.NY Other specified deletions of chromosome 22
LD44.NZ Deletions of chromosome 22, unspecified
LD44.P Deletions with other complex rearrangements
LD44.Y Other specified deletions of the autosomes
LD44.Z Deletions of the autosomes, unspecified
LD45 Uniparental disomies
LD45.0 Uniparental disomies of maternal origin
LD45.1 Uniparental disomies of paternal origin
LD45.Y Other specified uniparental disomies
LD45.Z Uniparental disomies, unspecified
LD46 Imprinting errors
LD46.0 Maternal imprinting error

20 Developmental anomalies
LD46.1 Paternal imprinting error
LD46.Y Other specified imprinting errors
LD46.Z Imprinting errors, unspecified
LD47 Balanced rearrangements or structural rearrangements
LD47.0 Balanced translocation and insertion in normal individual
LD47.1 Chromosome inversion in normal individual
LD47.2 Balanced autosomal rearrangement in abnormal individual
LD47.3 Balanced sex or autosomal rearrangement in abnormal individual
LD47.4 Autosomal fragile site
LD47.Y Other specified balanced rearrangements or structural rearrangements
LD47.Z Balanced rearrangements or structural rearrangements, unspecified
Sex chromosome anomalies
LD50 Number anomalies of chromosome X
LD50.0 Turner syndrome
LD50.00 Karyotype 45, X
LD50.01 Karyotype 46, X iso Xq
LD50.02 Karyotype 46, X with abnormal sex chromosome, except iso Xq
LD50.03 Mosaicism, 45, X, 46, XX or XY
LD50.04 Mosaicism, 45, X or other cell line with abnormal sex chromosome
LD50.1 Karyotype 47,XXX
LD50.2 Mosaicism, lines with various numbers of X chromosomes
LD50.3 Klinefelter syndrome
LD50.30 Klinefelter syndrome with karyotype 47,XXY, regular
LD50.31 Klinefelter syndrome, male with more than two X chromosomes
LD50.3Y Other specified Klinefelter syndrome
LD50.Y Other specified number anomalies of chromosome X
LD50.Z Number anomalies of chromosome X, unspecified
LD51 Structural anomalies of chromosome X, excluding Turner syndrome
LD52 Number anomalies of chromosome Y
LD52.0 Male with 46,XX karyotype
LD52.1 Male with double or multiple Y
LD52.Y Other specified number anomalies of chromosome Y
LD52.Z Number anomalies of chromosome Y, unspecified
LD53 Structural anomalies of chromosome Y
LD54 Male with sex chromosome mosaicism
LD55 Fragile X chromosome
LD56 Chimaera 46, XX, 46, XY
LD5Y Other specified sex chromosome anomalies
LD5Z Sex chromosome anomalies, unspecified
LD7Y Other specified chromosomal anomalies, excluding gene mutations
LD7Z Chromosomal anomalies, excluding gene mutations, unspecified
LD90 Conditions with disorders of intellectual development as a relevant clinical feature
LD90.0 Angelman syndrome
LD90.1 Early-onset parkinsonism - intellectual deficit
LD90.2 Pelizaeus-Merzbacher-like disease
LD90.3 Prader-Willi syndrome
LD90.4 Rett syndrome
5C55.01 Lesch-Nyhan syndrome
LA04.0 Hydrocephalus with stenosis of the aqueduct of Sylvius
8A44.0 Pelizaeus-Merzbacher disease
LD90.2 Pelizaeus-Merzbacher-like disease

20 Developmental anomalies
8C21.2 Hereditary sensory and autonomic neuropathy type IV
LD20.00 Joubert syndrome
5C50.0 Phenylketonuria
5C50.00 Classical phenylketonuria
5C50.01 Nonclassical phenylketonuria
5C50.02 Embryofetopathy due to maternal phenylketonuria
5C50.0Y Other specified phenylketonuria
5C50.0Z Phenylketonuria, unspecified
5C50.12 Tyrosinaemia type 2
5C50.A1 Carbamoylphosphate synthetase deficiency
5C50.F1 Carnosinaemia
5C50.F2 Homocarnosinosis
LD20.1 Syndromes with lissencephaly as a major feature
5C52.03 Sjögren-Larsson syndrome
LA05.50 Polymicrogyria
LA05.60 Porencephaly
5C53.02 Pyruvate dehydrogenase complex deficiency
CB04.5 Brain-lung-thyroid syndrome
5C56.02 Metachromatic leukodystrophy
5C56.1 Neuronal ceroid lipofuscinosis
5C56.31 Mucopolysaccharidosis type 2
5C56.33 Mucopolysaccharidosis type 6
5C60.0 Oculocerebrorenal syndrome
LD44.N0 CATCH 22 phenotype
LD24.80 Langer-Giedion syndrome
5C58.00 Crigler-Najjar syndrome
LD55 Fragile X chromosome
LD27.00 Incontinentia pigmenti
LD2D.2 Tuberous sclerosis
LD2F.15 Noonan syndrome
KA62.8 Congenital rubella syndrome
KA62.3 Congenital cytomegalovirus infection
LD40.0 Complete trisomy 21
LD50.31 Klinefelter syndrome, male with more than two X chromosomes
LD90.Y Other specified conditions with disorders of intellectual development as a relevant clinical feature
LD90.Z Conditions with disorders of intellectual development as a relevant clinical feature, unspecified
LD9Y Other specified developmental anomalies
LD9Z Developmental anomalies, unspecified

Appendix B: ICBDSR Suggested Initial List of Congenital Anomalies

Suggested initial list of congenital anomalies to consider for monitoring and relevant ICD-10 codes
Congenital malformations of the nervous system
Anencephaly Q00.0
Craniorachischisis Q00.1
Iniencephaly Q00.2
Encephalocele Q01.0–Q01.2, Q01.8, Q01.9
Spina bifida Q05.0–Q05.9
Cleft lip and cleft palate

Suggested initial list of congenital anomalies to consider for monitoring and relevant ICD-10 codes
Cleft palate Q35, Q35.1, Q35.3, Q35.5, Q35.59, Q35.9, Q87.0
Cleft lip Q36, Q36.0, Q36.9, Q36.90, Q36.99
Cleft palate with cleft lip Q37, Q37.0, Q37.1, Q37.2, Q37.3, Q37.4, Q37.5, Q37.8, Q37.9, Q37.99
Congenital malformations of genital organs
Hypospadias Q54.0–Q54.4, Q54.8, Q54.9
Congenital malformations and deformations of the musculoskeletal system
Talipes equinovarus Q66.0, Q66.1, Q66.4, Q66.8
Reduction defects of upper limb Q71.0–Q71.6, Q71.8, Q71.9
Reduction defects of lower limb Q72.0–Q72.9
Reduction defects of unspecified limb Q73.0, Q73.1, Q73.8
Abdominal wall defect
Exomphalos/omphalocele Q79.2
Gastroschisis Q79.3
Eye
Congenital ectropion Q10.1
Congenital entropion Q10.2
Absence of eyelashes Q10.3
Dystopia canthorum Q10.3
Epicanthal folds Q10.3
Epicanthus inversus Q10.3
Fused eyelids Q10.3
Long palpebral fissure(s) Q10.3
Upward or downward slanting palpebral fissures Q10.3
Short palpebral fissures Q10.3
Long eyelashes Q10.3
Weakness of eyelids Q10.3
Stenosis or stricture of lacrimal duct Q10.5
Coloboma of iris Q13.0
Brushfield spots Q13.2
Iris freckles Q13.2
Blue sclera Q13.5
Exophthalmos Q15.8
Strabismus Q15.8
Ear
Accessory tragus Q17.0
Auricular tag or pit Q17.0
Double ear lobule Q17.0
Ear pit or tag Q17.0
Preauricular appendage, tag or lobule Q17.0
Large ears Q17.1
Macrotia Q17.1
Absent tragus Q17.3
Asymmetric sized ears Q17.3
Crumpled ears Q17.3
Cup ear Q17.3
Ear lobe crease Q17.3
Ear lobe notch Q17.3
Lack of helical fold Q17.3
Lop ear Q17.3
Misshapen ears Q17.3
Pointed ear Q17.3

Suggested initial list of congenital anomalies to consider for monitoring and relevant ICD-10 codes
Primitive shape of ear Q17.3
Protruding ears Q17.3
Simple ear Q17.3
Small ears (excludes true microtia) Q17.3
Thickened or overfolded helix Q17.3
Low-set ears Q17.4
Misplaced ear Q17.4
Posteriorly rotated ears Q17.4
Bat ear Q17.5
Prominent ears Q17.5
Darwinian tubercle Q17.8
Narrow external auditory meatus Q17.8
Face and neck
Branchial vestige Q18.0
Branchial tag or pit Q18.0
Fistula of auricle, congenital and fistula cervicoaural Q18.1
Pretragal (commonly referred to as preauricular) sinus and cyst Q18.1
Redundant neck folds Q18.3
Webbed neck (pterygium colli) Q18.3
Macrostomia Q18.4
Microstomia Q18.5
Hypertrophy of lip, congenital or macrocheilia or large wide lips Q18.6
Short or long columella Q18.8
Angular lip pits Q18.8
Small lips Q18.8
Short neck Q18.8
Thin vermillion border Q18.8
Synophrys, confluent or medial flare eyebrows Q18.80
Peripheral vascular system
Single umbilical artery Q27.0
Nose
Notched or hypoplastic alae nasi Q30.2
Anteverted nares Q30.8
Flat or wide nasal bridge Q30.8
Small nares Q30.8
Smooth philtrum Q30.8
Mouth
Cleft uvula Q35.7
Tongue tie Q38.1
Macroglossia Q38.2
Adhesion of tongue Q38.3
Bifid tongue Q38.3
Fissure of tongue Q38.3
Hypoglossia Q38.3
Hypoplasia of tongue Q38.3
Microglossia (hypoplasia of tongue) Q38.3
Ranula Q38.4
Absent uvula Q38.5
High arched palate Q38.50
Aberrant frenula Q38.6
Broad alveolar ridge Q38.6

Suggested initial list of congenital anomalies to consider for monitoring and relevant ICD-10 codes
Cleft gum (in the absence of cleft lip) Q38.6
Natal teeth Q38.6
Anus and genitalia
Anterior anus (ectopic anus) Q43.5
Imperforate hymen Q52.3
Embryonal cyst of vagina Q52.4
Fusion of labia Q52.5
Fusion of vulva Q52.5
Prominent clitoris Q52.6 or Q52.8
Hypoplastic labia majora Q52.8
Hypoplastic labia minora Q52.8
Undescended testicle, unilateral Q53.1
Undescended testicle, bilateral Q53.2
Undescended testicle, unspecified Q53.9
Chordee (without hypospadias) Q54.4
Hypoplasia of testis and scrotum Q55.1
Shawl scrotum Q55.2
Retractile testis Q55.20
Bifid scrotum Q55.21
Absent or hooded foreskin of penis Q55.6
Curvature of penis lateral Q55.6
Phimosis Q55.6
Redundant foreskin Q55.6
Small penis (unless documented as micropenis) Q55.6
Hydrocele of testis Q55.8
Scrotalization of phallus Q55.9
Foot
Metatarsus varus or metatarsus adductus Q66.2
Hallux varus Q66.3
Congenital pes planus Q66.5
Hallux valgus Q66.6
Metatarsus valgus Q66.6
Pes cavus Q66.7
Hammer toe, congenital Q66.8
Long toes Q66.8
Prominent calcaneus Q66.8
Prominent heel Q66.8
Short great toe Q66.8
Vertical talus Q66.8
Widely spaced first and second toes Q66.8
Recessed fourth and fifth toes Q66.8
Short fourth metatarsus Q66.8
Short or broad great toe Q66.8
Rocker-bottom feet Q66.8
Overlapping toe Q66.8
Head, face, spine and chest
Facial asymmetry Q67.0
Compression facies Q67.1
Dolichocephaly Q67.2
Flat occiput Q67.3
Head asymmetry Q67.3

Suggested initial list of congenital anomalies to consider for monitoring and relevant ICD-10 codes
Plagiocephaly Q67.3
Squashed or bent nose, congenital Q67.4
Deviation of nasal septum Q67.41
Funnel chest Q67.6
Pectus excavatum Q67.6
Congenital pigeon chest Q67.7
Pectus carinatum Q67.7
Barrel chest Q67.8
Deformed chest Q67.8
Prominent sternum Q67.8
Shield-like chest Q67.8
Other musculoskeletal (including limbs)
Congenital deformity of sternocleidomastoid muscle Q68.0
Contracture of sternocleidomastoid (muscle) Q68.0
Congenital torticollis Q68.0
Camptodactyly Q68.1
Congenital clubfinger Q68.1
Long fingers Q68.1
Overlapping digits, not otherwise specified Q68.1
Short fourth metacarpal Q68.1
Single crease fifth finger Q68.1
Tapered fingers Q68.1
Short fingers Q68.1
Clinodactyly Q68.10
Genu recurvatum Q68.21
Cubitus valgus Q68.8
Hyperextended joints, not otherwise specified Q68.8
Hyperextended knee Q68.8
Polydactyly type B of fingers (type B is, by definition, post axial and rudimentary (postminimi); type A is post axial, fully developed is a major anomaly – Q69.02A – and is not a minor anomaly) Q69.02B
Polydactyly type B, not otherwise specified Q69.02B
Polydactyly type B of toes Q69.22B
Syndactyly (involving second and third toes) Q70.3
Genu valgum Q74.1
Other anomalies of skull, face and spine
Scaphocephaly Q75.0
Trigonocephaly, other head deformations without synostosis Q75.0
Hypertelorism Q75.2
Macrocephaly (includes familial benign macrocephaly) Q75.3
Hypotelorism Q75.8
Maxillary hypoplasia or prominence Q75.8
Micrognathia Q75.8
Prognathia Q75.8
Frontal bossing Q75.8 or Q75.80
Large or small fontanelles Q75.8 or Q75.80
Metopic suture open to bregma Q75.8 or Q75.80
Narrow bifrontal diameter Q75.8 or Q75.80
Prominent occiput Q75.8 or Q75.80
Prominent or hypoplastic supraorbital ridges Q75.8 or Q75.80
Third fontanelle Q75.8 or Q75.80
Spina bifida occulta Q76.0

Suggested initial list of congenital anomalies to consider for monitoring and relevant ICD-10 codes
Congenital lordosis, postural Q76.43
Abdomen
Diastasis recti Q79.5
Inguinal hernia Q79.8
Umbilical hernia Q79.8
Skin, breast and other integument
Skin cyst Q82.4 or Q84.4
Angioma Q82.5
Benign skin neoplasm, pigmented naevus (ear and auditory canal) Q82.5
Benign skin neoplasm, pigmented naevus (eyelid) Q82.5
Benign skin neoplasm, pigmented naevus (face) Q82.5
Benign skin neoplasm, pigmented naevus (lip) Q82.5
Benign skin neoplasm, pigmented naevus (lower limb, hip) Q82.5
Benign skin neoplasm, pigmented naevus (other specified site) Q82.5
Benign skin neoplasm, pigmented naevus (scalp, neck) Q82.5
Benign skin neoplasm, pigmented naevus (trunk) Q82.5
Benign skin neoplasm, pigmented naevus (unspecified site) Q82.5
Benign skin neoplasm, pigmented naevus (upper limb, shoulder) Q82.5
Cafe-au-lait spot Q82.5
Hemangioma (other than face and neck) Q82.5
Lymphangioma Q82.5
Noncavernous, single, small hemangioma (4 cm diameter) Q82.5
Pigmented naevus, congenital non-neoplastic naevus Q82.5
Birthmark Q82.50
Naevus flammeus Q82.51
Port-wine stain Q82.51
Strawberry naevus Q82.51
Mongolian spot Q82.52
Cutis marmorata Q82.8 or Q84.8
Dimple, hand Q82.8 or Q84.8
Dimple, shoulder Q82.8 or Q84.8
Extra or absent hand or interphalangeal creases Q82.8 or Q84.8
Pilonidal or sacral dimple Q82.8 or Q84.8
Plantar furrow Q82.8 or Q84.8
Rectal fissure Q82.8 or Q84.8
Sole crease Q82.8 or Q84.8
Vaginal or hymenal tags Q82.8 or Q84.8
Single transverse palmar crease Q82.80
Anal tag Q82.81
Skin tag Q82.81
Unusual dermatoglyphics Q82.84
Absent nipple Q83.2
Extra nipples (supernumerary nipples) Q83.3
Supernumerary nipple Q83.3
Inverted nipples Q83.8
Small nipple (hypoplastic) Q83.8
Widely spaced nipples Q83.8
Monilethrix Q84.1
Pili annulati Q84.1
Pili torti Q84.1
Aberrant scalp hair patterning Q84.1 or Q84.2

Suggested initial list of congenital anomalies to consider for monitoring and relevant ICD-10 codes
Depigmentary hair changes Q84.1 or Q84.2
Hair upsweep Q84.1 or Q84.2
Low posterior hairline Q84.1 or Q84.2
Persistent lanugo Q84.2
Congenital hypertrichosis Q84.20
Absent nails (major if third phalanx is missing) Q84.3
Enlarged or hypertrophic nails Q84.5
Pachyonychia Q84.5
Congenital clubnail, koilonychia, malformation of nail, not otherwise specified Q84.6
Duplication of thumbnail Q84.6
Hyperconvex fingernails Q84.6
Hyperconvex toenails Q84.6
Hypoplastic fingernails Q84.6
Hypoplastic toenails Q84.6
Thickened toenails Q84.6
Aplasia cutis (major if large) Q84.8

Appendix C: List of Congenital Anomalies Reported by EUROCAT for Prevalence Data

Nervous system anomalies
Neural tube defects
Anencephaly and similar
Encephalocele and meningocele
Spina bifida
Hydrocephaly
Severe microcephaly
Arhinencephaly/holoprosencephaly
Agenesis of corpus callosum
Eye anomalies
Anophthalmos/microphthalmos
Anophthalmos
Congenital cataract
Congenital glaucoma
Ear, face and neck anomalies
Anotia and atresia/stenosis/stricture of external auditory canal
Congenital heart defects
Severe congenital heart defects
Common arterial truncus
Double outlet right ventricle
Double outlet left ventricle
Complete transposition of great arteries (D-TGA)
Single ventricle
Corrected transposition of great arteries (L-TGA)
Ventricular septal defect
Atrial septal defect
Atrioventricular septal defect
Tetralogy and pentalogy of Fallot
Tricuspid atresia and stenosis

Ebstein's anomaly
Pulmonary valve stenosis
Pulmonary valve atresia
Aortic valve atresia/stenosis
Mitral valve atresia/stenosis
Hypoplastic left hear (HLH/HLHS)
Hypoplastic right heart (HRH/HRHS)
Coarctation of aorta
Aortic atresia/interrupted aortic arch
Total anomalous pulmonary venous return
PDA as only CHD in term infants
Respiratory anomalies
Choanal stenosis or atresia
Congenital pulmonary airway malformations (CPAM)
Oro-facial clefts
Cleft lip with or without cleft palate
Cleft palate
Gastro-intestinal anomalies
Oesophageal atresia with or without trachea-oesophageal fistula
Duodenal atresia or stenosis
Atresia or stenosis of other parts of small intestine
Ano-rectal atresia and/or stenosis
Hirschsprung's disease
Atresia of bile ducts
Annular pancreas
Anomalies of intestinal fixation
Diaphragmatic hernia
Abdominal wall defects
Gastroschisis
Omphalocele
Congenital anomalies of kidney and urinary tract (CAKUT)
Unilateral renal agenesis
Bilateral renal agenesis including Potter sequence
Multicystic renal dysplasia
Congenital hydronephrosis including ureter obstruction
Lobulated, fused and horseshoe kidney and ectopic kidney
Bladder exstrophy and/or epispadia
Posterior urethral valve
Prune belly syndrome
Genital anomalies
Hypospadias
Indeterminate sex
Limb anomalies
Limb reduction defects (LRD)
Transverse LRD
Longitudinal preaxial LRD
Longitudinal postaxial LRD
Longitudinal central LRD
Intercalary LRD
Club foot (Talipes equinovarus)
Hip dislocation
Polydactyly

Syndactyly
Other anomalies/syndromes
Craniosynostosis
Congenital constriction bands/amniotic band sequence resulting in major malformations
Situs inversus
Conjoined twins
VATER/VACTERL association
Pierre Robin sequence
Caudal regression sequence
Sirenomelia
Septo-optic dysplasia
Vascular disruption anomalies
Laterality anomalies
Teratogenic syndromes resulting in major malformations
Valproate syndrome
Maternal infections resulting in major malformations
Genetic disorders
Skeletal dysplasias
Down syndrome/trisomy 21
Patau syndrome/trisomy 13
Edwards syndrome/trisomy 18
Turner syndrome
Triploidy and polyploidy

Appendix D: List of Congenital Anomalies Requested for Inclusion in the ICBDSR Database

International Clearinghouse for Birth Defects Surveillance and Research (ICBDSR)

Form 1 - Report on Selected Malformations

Surveillance System	

Year	

Live births (LB)	
Stillbirths (SB)	
Total births	
Number of terminations of pregnancy (ToP) for birth defects	

Birth Defects	Number of cases (*)		
	Live Births (LB)	Still Births (SB)	Termination of Pregnancy (ToP)
Anencephaly			
Spina bifida			
Encephalocele			
Microcephaly			
Holoprosencephaly			
Hydrocephaly			
Anophthalmos			
Microphthalmos			
Unspecified Anophthalmos/Microphthalmos			
Anotia			
Microtia			
Unspecified Anotia/Microtia			
Transposition of great vessels			
Tetralogy of Fallot			
Hypoplastic left heart syndrome			
Coarctation of aorta			
Choanal atresia, bilateral			
Cleft palate without cleft lip			
Cleft lip with or without cleft palate			
Oesophageal atresia/stenosis with or without fistula			
Small intestine atresia/stenosis			
Anorectal atresia/stenosis			
Undescended testis (36 weeks of gestation or later)			
Hypospadias			
Epispadias			
Indeterminate sex			
Renal agenesis			
Cystic kidney			
Bladder exstrophy			
Polydactyly, preaxial			
Total Limb reduction defects (include unspecified)			
Transverse			
Preaxial			
Postaxial			
Intercalary			
Mixed			
Unspecified			
Diaphragmatic hernia			
Omphalocele			
Gastroschisis			
Unspecified Omphalocele/Gastroschisis			
Prune belly sequence			
Trisomy 13			
Trisomy 18			
Down syndrome, all ages (include age unknown)			
<20			
20-24			
25-29			
30-34			
35-39			
40-44			
45+			
unknown			

Total Births by maternal age

(*) Please, no blank cells: 0 = no case reported; NR = data not reported or not available

Notes on the data

GPSR Compliance

The European Union's (EU) General Product Safety Regulation (GPSR) is a set of rules that requires consumer products to be safe and our obligations to ensure this.

If you have any concerns about our products, you can contact us on ProductSafety@springernature.com

In case Publisher is established outside the EU, the EU authorized representative is:

Springer Nature Customer Service Center GmbH
Europaplatz 3
69115 Heidelberg, Germany

Batch number: 10731385

Printed by Printforce, the Netherlands